OREGON GOLD

Oregon Gold

A HISTORY OF MINING FROM THE CIVIL WAR
INTO THE PROGRESSIVE ERA

William F. Willingham

Oregon State University Press Corvallis

Cataloging-in-publication information is available from the Library of Congress.
ISBN: 978-1-962645-47-8 paper; 978-1-962645-48-5 ebook

∞This paper meets the requirements of ANSI/NISO Z39.48-1992 (Permanence of Paper).

First published in 2025 by Oregon State University Press
Printed in the United States of America

Oregon State University Press
121 The Valley Library
Corvallis OR 97331-4501
541-737-3166 • fax 541-737-3170
www.osupress.oregonstate.edu

Oregon State University Press in Corvallis, Oregon, is located within the traditional homelands of the Marys River or Ampinefu Band of Kalapuya. Following the Willamette Valley Treaty of 1855, Kalapuya people were forcibly removed to reservations in Western Oregon. Today, living descendants of these people are a part of the Confederated Tribes of Grand Ronde Community of Oregon (grandronde.org) and the Confederated Tribes of the Siletz Indians (ctsi.nsn.us).

Contents

Acknowledgments

I began working on this book during the early days of the COVID-19 pandemic. At the time, most research archives and libraries were closed. Fortunately, I could access online collections of historical newspapers, mining journals, government agency records, published articles, and other sources covering precious metal mining in the American West. Once archives and libraries reopened, I made efficient use of their relevant collections, especially benefiting from my previous online research.

Many individuals at historical museums and societies aided my research as I crafted this book. I especially want to call out the support I received from the professionals at the Oregon Historical Society. I was fortunate to receive the Oregon Historical Society Sterling Research Fellowship, and Scott Daniels and Renato Rodriguez were especially helpful in guiding me through the relevant research collections at that institution. I also received exceptional editorial assistance from Eliza Canty-Jones, editor of the *Oregon Historical Quarterly*, and her staff, specifically Erin Brasell and Jenny DuVander, as I wrote an article for the *Quarterly* based on my mining research. My search for illustrations for this book was greatly aided by Kitman Kienzle of the Grant County Historical Museum, John Brockman of the Baker County Library, and the above-mentioned staff of the Oregon Historical Society.

Several individuals provided valuable information and read various early versions of chapters in this book. Kristi Negri, an editor on a previous book I wrote, gave me her insightful advice on my initial writing. I owe a special thanks to Don Hann, retired Malheur National Forest Archaeologist, for sharing his vast knowledge of Chinese mining in eastern Oregon. He generously shared with me his historical materials collected over many years of field and documentary research. J. B. Bane, of the Sumpter Valley Railroad Restoration, combed their records for an answer to a specific question I had. One afternoon in Baker City, Tim and Kim Lethlean, owners of the Virtue mine, shared with me their extensive knowledge of Baker County mining history. Needless to say, none of these individuals is responsible for any errors in this book; those are mine alone.

Introduction

This book is conceived as a companion to my earlier work on eastern Oregon, *Starting Over: Community Building on the Eastern Oregon Frontier.* Whereas that book focused on the settling and development of an agricultural landscape, this work examines the progress of a newly opened mining frontier in a mountainous environment of the same region. The new volume can also stand alone as a study on a neglected topic concerning one aspect of the economic growth of Oregon in the late nineteenth and early twentieth centuries. Just as the lure of free land attracted homesteaders and livestock raisers to eastern Oregon in search of new opportunities, so did the prospect of riches from the scramble for gold and silver on the public domain. In both cases, the people involved had to construct communities or establish business concerns to accomplish their individual goals and evolve ways to successfully interact with the larger world. The main narrative begins with the discovery of gold in 1862 at Auburn and Canyon City in eastern Oregon and ends in 1910 with the decline of large-scale lode mining in the region.

This study describes and analyzes the impact of pursuing and producing precious metals, mainly in eastern Oregon. A more complete understanding of the consequences of the quest for mineral wealth also requires placing the story in a comparative context of such efforts elsewhere in the West. In the beginning, an important hallmark of mining communities in the West was that the people in those settings experienced physical isolation because of poor transportation and communication facilities, at least in the early years. How did the early miners cope with their remote circumstances? Their social interactions were intensely face-to-face, often leading them to form a set of shared values, mutual obligations, and a strong sense of place—the basis of community formation. The question arose, though, whether such relationships could build a lasting community once the precious metals ran out. This study attempts to investigate that query.

Finally, another basic condition of mining's existence was that all work or activities initially were governed by the natural rhythms of the changing seasons, length of daylight, and dependence on human and animal labor. Placer mining was especially dependent on the ready availability of sufficient amounts of running water. Seasonal fluctuations of water or drought often limited such mining operations. In addition, lode mining underground brought its own difficult ore recovery issues to overcome. In time, however, mining companies acquired and employed mechanical methods, scientific expertise, and financial capital to break free from the constraints imposed by the original environmental conditions. The employment of increasingly sophisticated mining methods and financial services served as the opening wedge of the modern industrial future. This late nineteenth- and early twentieth-century period represented the transition from a rural, preindustrial economy and society to an urban, industrial one. This study traces that modernization process through the lens of eastern Oregon mining communities.

The discovery of gold in the Blue Mountains of eastern Oregon in the early 1860s jump-started the economic growth and political development of Oregon. In particular, the recovery of stream-laden gold jolted the economy of Portland to new heights. The newfound wealth, moreover, led to the expansion of Oregon's transportation infrastructure—both on land and water—quickened the permanent settlement of the state's eastern section, created the need for basic local political structures, and sped the displacement and segregation of the Indigenous peoples of the region. Economic and political activities associated with the recovery of gold helped propel the continuing growth of the new state in the late nineteenth century. Gold mining in eastern Oregon also attracted significant numbers of Chinese miners who contributed greatly to the economic growth and ethnic diversity of the region and state.

Precious metal mining in eastern Oregon went through two phases between 1862 and 1910. The initial phase consisted primarily of placer or alluvial mining (1862–1885) while the second period (1885–1910) focused on lode or underground mining. Placer mining of surface deposits was chiefly the work of individuals or small groups needing little capital. Underground lode mining, however, required improved transportation systems to bring in heavy machinery such as stamp mills and drilling equipment; new techniques for tunnelling, recovering, and processing the metal; and large infusions of capital to finance the entire effort. It should be noted that the impact of Oregon's precious metal recovery has had only cursory coverage from mining historians, with some even doubting its importance in western mining history. For example, Rodman Paul, a leading Western historian of mining, wrote, "Eastern Oregon . . . [mining was] not as important as Montana and Idaho either initially or subsequently."[1] It is true that between 1870 and 1910, government reports never ranked Oregon higher than seventh in total production of gold among the precious metal mining states. Gold mining in Oregon, however, represented a significant factor in the growth of Portland and the eastern portion of the state prior to 1880. Only in the last twenty years of the nineteenth century did Oregon's agricultural sector produce large quantities of exportable grain and did salmon fisheries become productive. The lumber industry took off only after 1900. The arrival of the transcontinental railroad in the early 1880s would also further stimulate the state's economy, notably enabling the development of lode mining.

To create a profile of the mining communities in eastern Oregon, I have focused on case studies of the Granite Creek and the Elk Creek (Susanville) Mining Districts located in Grant County, Oregon, between 1862 and 1910. Granite and Elk Creek Districts represented the two methods of gold mining during the period of study: placer (eroded gold deposits in streambeds) and lode or quartz (hard rock, underground). I have sought to delineate these districts' chief ethnic, demographic, and economic characteristics. Who were these miners, where did they come from, and how mobile were they? What kind of landscape did they find upon arrival, and did they alter it through their mining processes? What sort of social, political, and economic order and institutions developed, and did these arrangements change over time? What was the relationship of these mining communities with the outside world, and how did these associations affect the economic and social viability of eastern Oregon and the state?

To answer these and other pertinent questions about the creation and evolution of eastern Oregon mining communities over time, I have analyzed a variety of documentary sources. These sources included federal decennial census schedules, federal and county land records, court records, government reports, municipal records, vital records, newspapers and magazines, personal correspondence, diaries, business journals, oral and written interviews, maps, photographs, archaeological reports, and relevant secondary literature. Where appropriate, I analyzed the data quantitatively, but I made every effort to maintain the focus on the relationships among people and places. Quantification also provided a firmer grounding for determining the usefulness of anecdotal information in establishing what was typical or atypical in the mining experience. Period commentators proved to be quite adept at promoting or overstating the wealth awaiting those interested in mining or investing in the gold fields of eastern Oregon.

Another reason for choosing Granite and Susanville for intensive study stemmed from the large numbers of Chinese immigrants in their populations between 1866 and 1890. The Chinese had a significant presence in Grant County throughout the late nineteenth century. In 1870, for example, they made up 42 percent of the total population of the county. Chinese miners, usually operating as small, independent companies, worked claims or mining rights they bought or leased from Euro American miners. By the early 1900s, returns from such mining efforts fell off, and most Chinese miners left the county to seek work elsewhere. A few, however, remained to take service jobs, such as cooks, laborers, domestics, laundrymen, or sheepherders. A handful even operated general merchandise stores in locations such as Susanville, Granite, and John Day. Fortunately, a good deal of archaeological field work has been done on the Chinese mining camps located in the Granite and Elk Creek Mining Districts, allowing a description of their mining methods and contribution to the settlement and growth of the region.

This study also describes the legal and technological context for understanding late nineteenth- and early twentieth-century precious metal mining. Congress's Mining Law of 1872 established the basic legal rights of miners and the process for claiming public lands for mining purposes, but the law's general terms meant that the courts had to interpret the precise meaning of the act. Miners and mining companies spent a great deal of money and time trying to resolve difficult, real-world questions concerning the often-contested ownership of mining properties. During the years covered by this study, many changes in mining equipment and processes occurred, and I have described and assessed how these shifts affected the operations and profitability of the mining industry in eastern Oregon.

A word about newspapers as a source for local history is necessary since I made considerable use of them in this study. While imperfect, they are probably the best available means for recapturing a community's life, texture, and economic activity over time. On a weekly and sometimes daily basis, local newspapers across America covered people's lives, movements, work habits, and property transactions; recounted a wide range of political, economic, and social activities; and revealed the values and beliefs of ordinary citizens. Scandal and crime, as well as uplift and success, got their due in the local press. Often in colorful language, local papers could rarely be accused of being dull. Unfortunately, full runs of most weekly newspapers rarely survived. Ephemeral by nature, they have suffered loss over time through fires, changes of ownership, and general neglect. Fortunately, newspapers often commented on the happenings in neighboring communities and printed correspondent news from those locales. I have systematically examined the newspapers of eastern Oregon and elsewhere that covered the mining news of the region.

Additionally, I found the San Francisco-based *Mining and Scientific Press* and the Portland-founded *Pacific Miner* especially helpful. The *Mining and Scientific Press* was a weekly newspaper covering mining throughout the American West between 1860 and 1922. It included substantial reportage on Oregon mining. The *Pacific Miner* appeared bi-monthly from 1899 to 1906. Both avoided sensationalism in their news and editorial columns—something many newspapers of the time could not claim—and provided solid coverage of technical advances in mining, legal opinions affecting the ownership and operation of mines, and other important facets of precious metal mining.

Even though prospectors discovered gold on the west side of the Cascade Mountains in southern Oregon between 1850 and 1851 and carried out placer and hydraulic mining there into the twentieth century, the production was not well-recorded by official governmental entities, especially in the early years. Jacksonville was the center of the 1850s gold rush, and later, mining flourished in the region north and east of Grants Pass. In addition, some limited lode mining occurred after 1890 in both Jackson and Josephine counties. The best compilation of mining statistics for the period between 1880 and 1899 found that Jackson and Josephine counties accounted for at most 22 percent of Oregon's gold and silver production. Since the total results of mining in southern Oregon apparently never came close to the quantities recovered in eastern Oregon, I have not provided in-depth coverage of that part of Oregon's embrace of golden dreams.[2]

This study frequently references the monetary value of gold and silver production and its economic impacts in Oregon between 1862 and 1910. During this entire era, the federal government set the official price of gold at $20.67 per ounce. Across this period, moreover, the dollar had a relatively stable value. To give the modern reader a sense of the historical dollars' worth in terms of today's dollars, I have employed an inflation calculator. Of the existing calculators, the one used here is the DaveManuel.com Inflation Calculator because it gives more conservative results compared to other calculators (i.e., the lowest compared amount). This calculator is based on Oregon State University's historical inflation data. To convert the historical dollar to a modern value, the reader should multiply by the average inflation factor of about 26.24 for the period between 1870 and 1910. Thus, $1,000,000 becomes $26,240,000 in today's money. For the 1860s, the inflation factor is slightly less: 18.11. The narrative generally provides just the historical dollar amount for money during the period under study. The reader can readily apply the inflation factor for conversion to modern values.[3]

Another way to understand the value of the historical dollar involves examining Oregon's wages and cost of living in the past. In general, wages and prices were relatively stable in the second half of the nineteenth century.

A common laborer, for example, typically earned $1.50 to $2.50 a day. A skilled laborer, on the other hand, could earn $3.50 to $5.00 a day. Because of the hazards of the job, miners usually earned between $3.00 and $5.00 a day. A six-day workweek (54 hours) was standard, but workers did not necessarily work a full year; periods of unemployment were common. On balance, typical yearly wages for a common laborer came to about $600 to $700. For added context on wages during the period under study, note that business and professional occupations earned higher incomes. Store clerks and bookkeepers, for example, were paid within a wide range, depending on importance: $30 to $100 a month. A civil engineer, depending on training and experience, could earn between $150 and $200 per month. A college professor might receive $900 to $1,500 a year, depending on rank, whereas the president of such an educational institution, depending on whether it was private or public, got paid $1,500 to as much as $3,000 a year. At the bottom end of the pay scale for an educator, a rural schoolteacher's pay typically came to about $45 a month, plus board and room for a six-month term. A ranch hand also earned about $25 to $40 per month, with board and room provided.[4]

Information concerning consumer prices for everyday items provides further context for assessing the standard of living in eastern Oregon's mining communities and of the region in general. What did an income buy in a mining community? When assessing living standards during the period between 1870 and 1910, it is important to note that the tax burden was comparatively light for most income levels. Daily wage earners or monthly salaried personnel for all types of employment kept most of what they made. For much of the period, state and federal governments did not impose an income tax, and state and local taxes on real and certain types of personal property were usually low. What did the common items of living cost? The 1902 edition of the Sears, Roebuck Catalogue provides some answers. In terms of clothing, a man's wool suit cost from $5 to $10; a woman's dress, $3.75 to $6.75. Men's leather shoes could be had for $2.85, and men's overalls, 50 to 75 cents. Boots for miners cost $1.50 and for loggers, $2.40. Cast-iron cook stoves sold for between $15 and $25, and a sewing machine for $12.85. A man's typical pocket watch could be had for $2, and a cast-iron tea kettle cost $1. A luxury item, such as a white enameled bathtub, sold for $24.50. A 100-piece set of crockery dinner dishes could cost between $5 and $9. Bicycles sold for $8.95 to $15.75. In Baker City, at the turn of the century, a five-room house rented for $10 to $15 per month. Hotels charged $1 to $2.50 a night, depending on quality and location. Groceries at that time (1902) in Baker City included the following prices for staples: Fresh beef, pork, and chicken sold from 10 cents to 20 cents per pound. Fresh vegetables, from 1 cent to 2 cents per pound.

The Sears Catalog listed five pounds of coffee for $1.10 and one pound of tea for 47 cents.

Oregon's early economic development stemmed from exploiting its abundant natural resources—agricultural land, timber, fisheries, and precious metals. Since all but the role of gold and silver have been written about in some detail, this study completes the story by focusing on the forgotten part. Indeed, between 1862 and 1910, precious metals ranked in the top five of the most valuable products of Oregon, and the state annually ranked between seventh and ninth nationally in the output of gold and silver. Additionally, placing the narrative and analysis in a larger western context captures what is unique or not about the Oregon mining experience.

Portland, Oregon, 1857. First Street was the setting for the gold seekers planning their 1861 expedition to eastern Oregon. Courtesy of the Oregon Historical Society (OHS) Research Library.

Chapter One
Gold Produces Astonishing Growth

In 1860, hardly any Oregonians lived outside of the Willamette Valley and southwestern part of the state. That year, the federal census recorded fewer than 100 inhabitants east of The Dalles. These hardy souls resided mainly in the Umatilla River and Grand Ronde River valleys of eastern Oregon. The discovery of gold at Griffin's Gulch in the fall of 1861, however, dramatically changed the situation. Soon, an estimated ten thousand gold seekers arrived, establishing communities such as Auburn, Canyon City, and Independence (Granite). By 1864, four new counties would be carved out of Wasco County to serve this new population. The riches flowing out of eastern Oregon and Idaho mines quickly supercharged the growth of Portland's commerce and finance. To accommodate the increased traffic to the gold fields, the federal government carried out improvements on the Columbia River and surveyed or subsidized overland routes in eastern Oregon, expanding the state's transportation infrastructure. The mining rush also brought significant numbers of Chinese people to eastern Oregon. Mining, however, had a less positive effect on the Indigenous peoples of the region, causing their displacement and the dispossession of their native lands.

GOLD FEVER

The initial discovery of gold in eastern Oregon came by accident. In August 1845, a group of Oregon-bound pioneers, consisting of 200 wagons and about 1,000 emigrants (the so-called "Lost Wagon Train"), left the Oregon Trail near present-day Vale and headed directly west in what they hoped was a shortcut to the Willamette Valley. Led by Stephen Meek, they unfortunately wandered aimlessly across the high desert country of the Great Basin. The party suffered at least twenty-three deaths before turning north along the Deschutes River, eventually reaching The Dalles in October 1845. At a stop along the arduous journey, some children playing in a stream, possibly a tributary of the John Day River (Wahawpan in Columbia River Sahaptin), found shiny pebbles and tossed them into a blue bucket, not realizing that they might be gold.[1] A few of the rocks were carried on to the Willamette Valley and determined to be gold, but they were subsequently lost. The discovery of gold in California in 1848 and southwestern Oregon in 1851 reawakened memories of the Meek party's possible discovery of gold, but no organized effort to search for the metal in eastern Oregon took place until 1861.[2]

In June of 1861, talk of gold strikes in the Pacific Northwest pervaded the air. The editor of the *Oregonian*, writing in response to mining discoveries in Washington Territory and British Columbia, described the attraction of gold to different groups of men:

> Besides those who are out [of] employment there is another class of miners: Those who have made money else where [*sic*], and whose prosperity has made them discontented with their slow gains in the old claim. These men are willing to rush off to the first excitement, and 'play to double or break.'

Another group, he wrote, included,

> [those] connected with agricultural employments, who have lands mortgaged or debts hanging over them, which they hope to liquidate speedily by a successful season in the mines.

The editor went on to warn,

> while all gold mining countries hold out inducements for the young and adventurous, only a small portion ever acquire a competency by mining. The troubles, the misfortunes of the unlucky miner have never been told; his weary marches, his coarse food, his scant apparel, his uninviting labors and his voluntary banishment from all the amenities of social intercourse. Those who raise mining excitements are generally sufficiently wide awake to avoid that mode of life, which they endeavor to stimulate others to adopt.[3]

Such advice went largely unheeded.

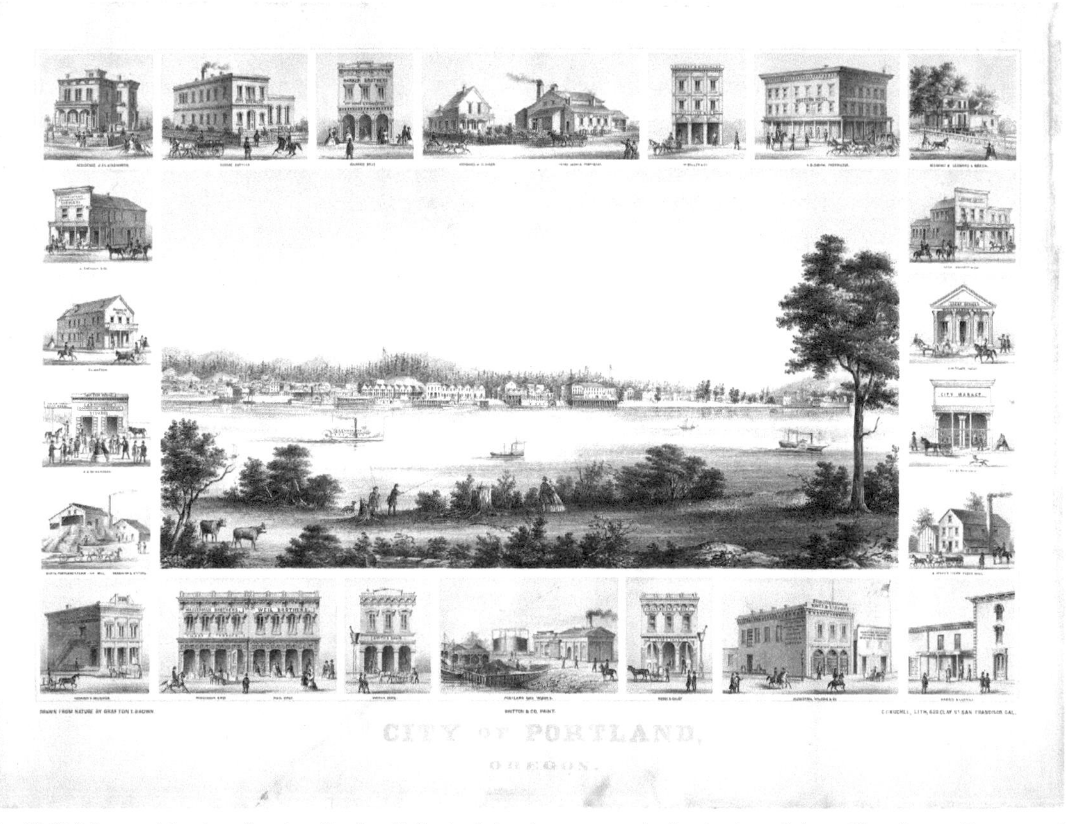

An 1863 lithographic view showing Portland's limited development at the beginning of the gold rush era. Courtesy of the Amon Carter Museum of American Art.

Shortly after that editorial appeared, a man by the name of I. L. Adams, claiming to have been a member of the 1845 "Lost Wagon Train," told a group of men gathered on Front Street in Portland that he knew of a stream with gold in eastern Oregon and offered to lead a party there. Among his listeners were four miners from the gold fields of California who had stopped in Portland on their way to the Oro Fino mines in Washington Territory: Henry Griffin, David Littlefield, William Stafford, and G. W. Schriver. By August, Adams and the four Californians had organized a well-equipped prospecting party of about 60 men to seek the location of Adams' alleged gold find. Once on the hunt, it soon became apparent that Adams had never been in the Meek Party of 1845 or had never known the route taken by those pioneers. After much aimless searching, the exhausted gold seekers encamped in the area between the headwaters of the Malheur and Burnt rivers and turned their wrath on Adams. Following heated discussions and a trial of sorts, some of the disgruntled prospectors called for lynching Adams. Cooler heads prevailed, however, and it was decided that Adams should be banished from their camp without supplies and left to find his own way back to Portland. Apparently, a few men, feeling sorry for Adams, gave him some supplies and aided his escape. By mid-October, most of the prospecting party decided to give up the search and return to their homes in the Willamette Valley over the same route taken on the outbound journey, but twenty-one others carried on the quest, heading northeasterly. The latter group planned to prospect until reaching the Oregon Trail and then return home over that road.[4]

GOLD DISCOVERED IN EASTERN OREGON

The discovery of gold in eastern Oregon in late 1861 was not the state's first encounter with the precious metal. Prospectors found placer gold near what is now Jacksonville in the southwestern part of Oregon in 1851 and 1852, and soon, Jackson County attracted large numbers of miners and settlers. A network of roads and coastal sailing operations connected the region with San Francisco, and placer gold flowed to the mint in that city. Unfortunately, accurate accounts of the gold recovered before the 1870s do not exist, but subsequent government records indicated that southwestern Oregon never produced more than 20 to 25 percent of the state's annual total between 1880 and 1910. In fact, once gold was discovered in eastern Oregon, many prospectors from Jackson and Josephine counties made the trek to the new mining area.[5]

On October 23, 1861, Henry Griffin and the other three Californians camped at a location on a tributary of the Powder River (Haawpala, in Cayuse Nez Perce) on the western slope of the Elkhorn Mountains (Tamapatuktpa, in Columbia River Sahaptin), subsequently called Griffin's Gulch. Before settling in for the night, Griffin decided to try his luck and sunk a shallow hole in the stream bank. Panning the recovered dirt, he found paying amounts of gold. After staking several claims in the gulch, the group traveled to Walla Walla, procured supplies, and returned to the gulch to spend the winter and continue the hunt for gold. A few of the party, not wanting to winter in the mountains, decided to return to Portland. Word of the discovery of gold soon spread. The *Oregonian* reported on November 9, 1861, "Most of the Adams prospecting party have got in . . . [and] are well satisfied with their prospect, and intend to go back in the spring." Of the original party of miners, eleven spent the winter and the following spring prospecting the territory around Griffin's initial discovery and uncovered other rich ground. They would soon be joined by a multitude of gold seekers from throughout the West.[6]

The search for richer soil resulted in a prospect four miles south of Griffin's Gulch in May 1862. Immediately, all the prospectors abandoned Griffin's Gulch for the new location. At this spot, the miners assembled on May 2, 1862, and drew up a set of mining laws for what they called the Blue Canyon District: The mining regulations defined the allowed size of claims,

Auburn, Oregon, residents showing placer mining equipment, ca. 1870. Courtesy of the OHS Research Library.

the work requirements necessary to retain the claim, the process for recording them and the methods for settling any disputes. Article Six clearly stated, "Chinamen shall not be allowed to own mining claims in this district, either by purchase or preemption." The final article called for choosing a recorder to "keep a book of records." In response, the miners next elected a recorder of claims to ensure the legal protection of their mining property. Without such quasi-legal protection, claim jumping was always a danger as more gold seekers arrived in the mountains of eastern Oregon. Within a year, the recorder registered over 1,200 claims. The creation of a mining district to limit claim size, protect claim ownership, mandate minimum work requirements to hold a claim, allocate water rights, and establish a means of adjudicating disputes was based on the practices growing out of the California gold rush experience.[7]

By June 1862, several hundred adventurers had spread over eastern Oregon in search of rich deposits of gold. They found enough paying dirt, consisting of flakes and nuggets, in the stream banks of one location that a permanent settlement soon sprang up. The miners laid out a townsite at the junction of Blue Canyon and Freezeout Creek—a short distance from

An early view of Auburn, Oregon, log houses set on piles to raise them above heavy winter snow. Courtesy of the OHS Research Library.

Griffin's Gulch—and named it Auburn. These prospectors had traveled from a variety of mining camps located throughout the Pacific Northwest, British Columbia, California, and Nevada, in hopes of finding more success than had been the case in their last venture.[8]

An early arrival among the throng of prospectors, Winfield Ebey, described in his journal the chaotic conditions that prevailed at the site of Auburn:

> We decend[ed] to "Griffin's Gulch," the first mines worked in this vicinity. We did not stop but pushed on a short distance &

> nooned. Then passed on by Elk Creek & to this place [Auburn] a couple of miles. At Elk Creek there is a good deal of mining going on, but we did not learn how the claims were paying. At this point there is quite a mining camp scattered some half mile along a small stream, Blue Canyon Creek. There is one log house. I do not know its use. The balance of the buildings are of canvass. There is a butcher shop kept on the flattened top of a big pine log. I believe there is a liquor saloon, but have not seen it. The miners seem busy at work & making something. They exhibited their gold. It is very course [coarse] & unwashed.[9]

Twelve days later, on June 16, Ebey noted in his journal the progress taking place at Auburn:

> We have a very pleasant location, & no doubt in the town—or where it will be soon, as it is the best site. Lots for building purposes are being taken up very rapidly. A man wishing a building lot squats on it, puts up his house & holds it. Between Freeze Out & Blue Gulches the town is regularly laid off & nearly all the lots taken up. Below the mouth of Freeze Out the lots [are] not surveyed, but every one is taking up lots for himself. Log houses are going up in every direction. Teams are hauling logs and axe is heard in every direction cutting timber for building purposes. Every thing is lively. . . . Ditches are being built to supply all who lack water & every thing betokens life, activity—and paying diggings. Nearly every claim in the vicinity are paying well & somc of them big.[10]

One month later, a correspondent to the Walla Walla *Washington Statesman* confirmed the astonishing growth of the settlement:

> Auburn is fast building up, and will now vie with any mining camp in the northern mines for useful improvements, considering its age. We have three store buildings, and several more in progress; two saloons, two blacksmith shops and one under construction; three butcher shops, one boarding house and several dwelling houses. Mr. Packwood and Company have sent for a portable sawmill, which they are quite sanguine will be here and in operation in the course of two months. . . . Provisions are plenty in the market at present. Whiskey rather plenty—judging from

> the debilitated appearance of several men now occupying, not a bed of down, but a bed of gravel along the sidewalk.[11]

The newspaper correspondent claimed that at least 2,000 persons resided in Auburn in the summer of 1862 and that probably 5,000 would be there by winter. Another visitor to Auburn in the fall of that year recorded in his diary that "the town which started early in the spring now contains as many as 1,000 houses and some 60 stores." At first, most shopkeepers operated their stores out of log homes. The notion of permanence, like so many gold-rush communities, was an afterthought in the early days of the search for gold. Auburn's mushroom growth was typical of other mining camps of the era.[12]

Auburn sat at the center of what proved to be a rich mining region. The district covered rugged, mountainous terrain, extending roughly eight by twelve miles. In this area, early mining was placer, where the miner recovered coarse gold by washing the dirt and debris excavated from stream banks and dry gulches in sluice boxes or simple pans with available water. Unfortunately, the supply of water was limited. At first, the miners dug ditches to bring water to Auburn from numerous small streams that drained into the nearby Powder River. When this was found to be insufficient, Portland capitalists organized a company between 1862 and 1863 to build a big ditch—the Auburn Canal—at a cost of $225,000 in historical dollar values.[13] According to W. H. Packwood, an investor in the company, this undertaking represented the only large-scale capital investment coming from Portland to the placer mines of eastern Oregon during the 1860s. When completed, the waterworks consisted of a canal extending thirty miles and a reservoir above the town of Auburn. The system distributed water to several gulches surrounding Auburn. The chief investors included Portland banker William Ladd, transportation czar Capt. John C. Ainsworth, and financier Simeon Reed. As major shareholders of the transportation monopoly, the Oregon Steam Navigation Company (OSN), they had a strong incentive for such a large investment.[14]

TRANSPORTATION TO THE GOLD FIELDS

The OSN, organized in 1860, controlled the most direct water route to the mining country of eastern Oregon and stood to make a lot of money transporting passengers and freight to and from the mines. The company's steamboats plied the Columbia and Snake rivers for 365 miles between Portland and Lewiston, Idaho. Its route also required using short rail portages (six miles) around the natural river obstructions at the Cascades of the Columbia and between The Dalles and Deschutes River (fifteen miles). The number of passengers and amount of freight and gold carried by the OSN during the height of the first boom in eastern Oregon and Idaho mining was striking:

Table 1.1

Year	Number of Passengers	Freight (tons)
1861	10,500	6,290
1862	24,500	14,550
1863	22,000	17,646
1864	36,000	21,834

These numbers take on meaning when one realizes that in the early 1860s, Portland had a population of about 3,500, and the entire state of Oregon, about 55,000. The company made a substantial profit from passengers, freight, bullion shipments, and a mail contract. It also benefited handsomely from a government contract to carry military supplies upriver. Steamboats earned fares totaling $1,000 to $6,000 a trip and sometimes as much as $11,000. One trip of the steamboat *Tenino* brought in $18,000 from freight, fares, and meals. By one estimate, using historical dollar values, OSN paid $4.6 million in dividends between 1860 and 1880. At the latter date, the owners sold the company for $5 million to Henry Villard, an eastern capitalist and railroad builder. By that time, OSN owned twenty-six vessels and had invested about $3 million in its facilities. Mining historian William Trimble noted, "Far-reaching enterprise, efficiency, and monopolistic grasp were, therefore, the outstanding characteristics of the Oregon Steam Navigation Company." As Jacob Kamm, one of the original owners, would succinctly state in 1910, "the Oregon Steam Navigation Company was the financial wonder of its day and age."[15]

Scene of the Oregon Steam Navigation Company's sidewheeler steamboat, *Oneonta*, meeting the train operating at the portage around the Cascade Rapids of the Columbia River in 1867. The *Oneonta* ran on the route between Portland and The Dalles. Courtesy of the OHS Research Library.

The OSN could not have operated so successfully without the assistance of the federal government. In its natural state, the Columbia was a treacherous, swift-flowing river with many rocky obstructions, in addition to the major interruptions at the Cascades of the Columbia and Celilo Falls. At low water periods, shallow draft steamboats traveled at their peril. Three rapids above The Dalles—Umatilla, John Day, and Homly—required the most work. To remedy the situation, the OSN and other commercial and development interests in Portland and eastern Oregon successfully lobbied Congress to appropriate funds for navigation improvements on the Columbia and Snake rivers. Responsibility for carrying out the arduous river work went to the US Army Corps of Engineers. From 1868 to 1918, the Corps expended about $700,000 to create a six-foot-deep river channel between Portland and Lewiston, Idaho, for shallow draft steamboats. In addition, between 1878 and 1895, the Corps labored to construct a canal and locks at the Cascades of the Columbia at a cost of $3.8 million. Next, it built The Dalles-Celilo Canal between 1905 and 1915 at a cost of $4.1 million.[16] The transformation of the river to accommodate the profit goals of miners, merchants, and settlers elided the fact that native peoples had flourished for millennia in harmony with the Columbia as nature had made it.

Once they reached The Dalles, OSN passengers had two choices for the overland route to the eastern Oregon mines. They could leave the steamboat at The Dalles, procure an outfit, and then trek another 250 miles over rough trails. Alternatively, they could travel further upriver to the Umatilla Landing (founded in 1863) and from there take one of several arduous paths another 150 miles through the Blue Mountains to the mines. Initially, the trails were too primitive to accommodate wagons. So most freight bound for the mines was carried by mule or horse packtrains. Most gold seekers simply walked to their destination because they could not afford to do otherwise. Gradually, as new mining grounds were discovered during the 1860s, entrepreneurs improved the overland routes. Stage lines began service out of Umatilla Landing in March 1864, with relay stations every ten to fifteen miles along the route. Still, since stage rides were expensive, most gold seekers continued to walk. A major incentive for investment in road improvements in eastern Oregon came from the continued discovery of gold deposits in the Blue Mountains from 1862 to 1864.[17]

CANYON CREEK GOLD RUSH

The next major gold strike occurred just as the Auburn miners began turning that community into a more permanent settlement. On June 7, 1862, a group of California prospectors—headed for the Idaho gold discoveries—camped on the banks of a stream that emptied into the main stem of the John Day River, two miles to the north. The next day, William Allard panned a large amount of gold, igniting a stampede to what became known as Canyon Creek. Other Californians soon crowded in. As one commentator later noted, "The spring of 1862 was a season of migration among the California miners, nomadic as were their habits, and ever ready as they were to take up the line of march whenever a promising discovery was made."[18] In a series of letters to his wife back in California, G. I. Hazeltine told of the hardships encountered by his party on the journey to the eastern Oregon mines and the early conditions along Canyon Creek. Hazeltine and nineteen compatriots (out of thirty-two originally) finally arrived at the mines after an arduous horseback trek of a month. They reached Canyon Creek on July 4, 1862, "tired, sore and nearly worn out, but in good health and spirits." He continued, optimistically, "Of the trip so far I shall not have space or time to tell you, suffice it to say that after going through all the privations, vexations and all the other actions except starvation that we are still alive and in hopes of doing well." In August, Hazeltine estimated that the mining camp had at least 1,000 miners and stated that "building is going on rapidly."[19]

Prospecting had already taken off at a rapid pace, and by June 25, 1862, the Canyon Creek miners assembled and adopted regulations governing the recording and working of all claims. The rules mirrored those adopted at Auburn, including the prohibition of Chinese miners. Over time, authorities would fail to enforce the ban as the Chinese, by 1870, would comprise close to 80 percent of the miners in the region (see chapter 6). The miners next laid out a town site along the narrow banks of Canyon Creek and named the settlement Canyon City. The miners aggressively carried on primitive placer mining, constructing two ditches, one two miles in length and the other ten miles long, to transport additional water to their claims. The miners' efforts even received notice in the San Francisco press through the September 11, 1862, issue of the *Mining and Scientific Press*: "The Miners on Canon Creek are working at a great disadvantage, owing to the lack of capital and the scarcity of lumber. Everything goes to show the richness of this creek and as soon as it is flumed and scientifically worked good results may be counted on."[20] Such news about gold strikes in distant northeastern Oregon helped send a steady stream of precious metal seekers northward from California.

In the early twentieth century, a government geologist estimated that the Canyon Creek miners of the 1860s recovered as much as $15 million during the first three years of effort and another $1.5 million a year over the next five years at a time when gold was valued at $20.67 an ounce. Another government mining expert reported that at the peak of placer mining in Canyon Creek during 1865, gold shipments averaged $22,000 a week. In all, miners probably recovered gold estimated at $26 million from a few miles' stretch of

Canyon Creek before the output tapered off after 1870. Still, the returns from the Canyon Creek placers averaged close to $100,000 a year up to 1900—much of it produced by Chinese mining companies. This was Oregon's richest gold placer operation.[21]

The crunch of thousands of miners led to the explosive growth of Canyon City, which sported the usual host of saloons, mercantile stores, boarding houses, livery stables, blacksmiths, and offices providing various professional services. As with Auburn, the earliest structures consisted of tents and log houses. Hazeltine wrote his wife that his house was "eighteen by twenty feet and eight logs high, [and] cost us about three hundred and fifty dollars, cash." Soon, though, enterprising businessmen brought in portable sawmills to both communities that supplied lumber for more substantial buildings. One difference between the communities of Auburn and Canyon City soon emerged. Canyon City sat next to the fertile John Day River valley, attracting homesteaders who started farming and livestock raising to supply the basic food needs of the miners in the vicinity. Auburn, located at a higher elevation and in rougher terrain, failed to develop such a non-mining economy. At their peak of mining excitement, each community is thought to have had from 3,000 to 5,000 inhabitants.[22]

ESTABLISHING LAW AND ORDER

With the influx of an estimated 10,000 people to eastern Oregon by the fall of 1862, the need for regular legal and administrative processes conducted by elected officials became apparent. The nearest seat of government where deeds could be recorded, licenses issued, and criminal and civil courts convened was The Dalles, which served as the county seat for a vast Wasco County, covering all of Oregon east of the Cascade Mountains. The need for instruments of law and order soon arose in the raw mining camps of Auburn and Canyon City. The first sign of criminal behavior came quickly in Auburn. In September 1862, the citizens there discovered that two men had been poisoned. Another man was held for the crime; but since the nearest seat of justice was 250 miles distant in The Dalles, the local citizens held a quasi-legal trial at Auburn. The jury convicted the accused man, and he was hanged. According to an eyewitness account, the entire community watched the spectacle of that violence: "I was grieved to notice at least fifty women, some with babes in their arms, and others mere girls, all desirous to see the dying throes of the doomed man." Two months later, in November 1862, a mob seized another man accused of fatally stabbing two acquaintances in a gambling game gone wrong and hanged him without a trial. After that event, an observer succinctly noted in his diary, "A Spaniard killed two men here with a knife. The miners took him and drug him down town by the heels and then hung him. A greaser shot into the crowd and wounded 3 men. The miners shot him." Auburnites were determined to have law and order, one way or another. The shipments of gold dust from the mining camps also encountered criminality, with highwaymen holding up private teamsters and expressmen.[23]

Citizens of Canyon City also experienced problems with law and order early in the town's existence. In the fall of 1862, Winfield Ebey wrote in his diary, "Canyon City is now infested with a set of men of the most desperate character. If the citizens don't administer a little justice to them soon, they will not perform their duty."[24] As with Auburn, Canyon City turned to quasi-legal procedures. In the spring of 1863, a miner, Berry Wey, allegedly killed his partner and stole his mule. After capturing the accused Wey and taking him to Canyon City, the locals held a "trial," convicted Wey, and hanged him the following day. Many years later, one early Canyon City miner reminisced that saloons were the location for "frequent rows and fights. There was neither law nor order . . . until a vigilance committee was organized sometime in the month of October [1862]. That organization soon suppressed the lawless element."[25] As western mining historian Paula Mitchell Marks has observed, "two types of criminal activity, personal physical attacks and robbery/theft, stand out on the gold rush frontier."[26]

To meet the need for law and order and convenient legal services, the state legislature in September 1862 carved two counties from the expansive area covered by Wasco County. One new county, named Umatilla, reached from Washington territory in the north to the North Fork of the John Day River in the south, while Willow Creek roughly defined the border with Wasco County on the west. The other new county, Baker, extended north and south between the Washington and Nevada territories and bounded Idaho territory on the east along the Snake River. The summit of the Blue Mountains served as the dividing line between the two new counties. Initially, Auburn served as the county seat for Baker County, while Umatilla Landing held the Umatilla County seat. Both locations proved to be temporary designations. As the gold rush populations decreased, the more permanent settlers in the two counties voted to move the seats of government to more central locations. In 1868, the new town of Pendleton gained the Umatilla County seat while upstart Baker City secured the Baker County government.[27]

Still, given the difficulties of travel, mostly on foot or horseback, these two new counties proved too large for the convenience of most residents who needed to conduct legal matters. The legislature once again accommodated them, in October 1864, by further subdividing Wasco County and then Baker County. Reflecting emerging settlement patterns, the northern section of Baker County became Union County, with the town of La Grande as county seat. The Powder River and its north fork served as the boundary line between the two counties. The entire southeastern portion of Wasco County became Grant County, with Canyon City designated as the county seat. The 45th parallel marked the north/south dividing line separating Umatilla and Grant counties, and the Nevada territory line marked the southern boundary of the new county. These rapid political changes indicated both the era's substantial population shifts in eastern Oregon and that population's influence in faraway Salem.

GOVERNMENT-SPONSORED TRANSPORTATION IMPROVEMENTS TO THE GOLD FIELDS

Roadbuilding in eastern Oregon before and during the early years of placer mining relied on investment from the federal government, which would benefit from the gold taken out of the region—wealth that was especially important while fighting the Civil War. The efforts achieved mixed results in transportation infrastructure but brought to non-Native people a significant new understanding of the region's landscape, knowledge that federal actors used to oppress Indigenous peoples and support miners and settlers.

Before the gold rush to eastern Oregon in the early 1860s, the soldiers stationed at Fort Vancouver investigated the possibilities of improving transportation routes in the region. The movement of emigrants over the Oregon Trail to the Willamette Valley had led the Army to send exploring parties to the region in hopes of finding shorter ways across its rugged terrain. While most civilians had given up finding a shortcut for the Oregon Trail through central and eastern Oregon, the Army had not. In the spring of 1859, the commander at Fort Vancouver, General William Harney, ordered Captain Henry Wallen to locate a wagon route through the region.

Wallen—and a well-equipped detachment of 193 soldiers, 154 horses, 344 mules, 121 oxen, thirty wagons, and an ambulance—marched south along the Deschutes River, crossing the Tygh Valley, and then proceeded southeasterly along the John Day River to its headwaters in the Blue Mountains. From this location, the party crossed the highest ridges to the Malheur River, descended that stream to its juncture with the Snake River, and at that point, made connection with the well-established emigrant trails. Partway through the reconnaissance, when the terrain became very rough, Wallen split his command, sending the heavy wagons back to The Dalles. With a reduced command of eighty-five men and a light pack train, he headed east to a large lake, named Lake Harney in honor of General Harney. This part of the mission demonstrated that a wagon road was possible to that point. Next, he headed northeast, encountering impassable terrain in the headwaters of the Malheur River and the southeastern edge of the Blue Mountains. His expedition finally crossed the Owyhee River and into southwestern Idaho. For

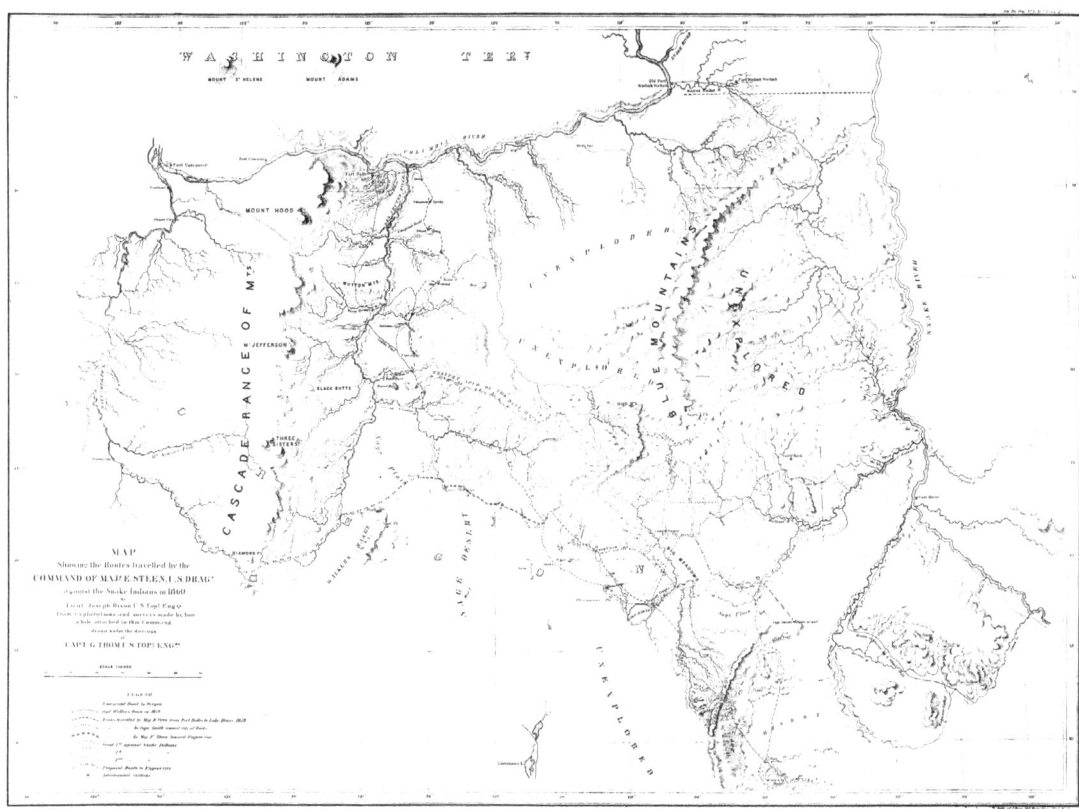

An 1860 US Army map showing most of Oregon east of the Cascade Mountains as "Unexplored." Courtesy of the author.

his efforts, Wallen admitted he had come up short: "I was somewhat disappointed in our route, as I expected to find it better than it really is. A wagon road cannot be constructed over a chain of mountains such as these before us without having hills to pull over; all the science of the engineer cannot change the general features of the country." He returned to The Dalles over the Oregon Trail. In 1860, another army expedition entered eastern Oregon under the command of Major Enoch Steen, set out to survey a road from Eugene to the California Trail in the Great Basin via Harney Lake. A possible route was found, but the advent of the Civil War prevented further attempts by the Army to lay out a wagon road in eastern Oregon.[28]

Merchants and outfitters in The Dalles, as well as Umatilla Landing, sought to capture the business of the miners swarming into eastern Oregon. To gain the trade at Canyon City, The Dalles merchants sponsored improvements to the trails that reached Canyon City via the central and eastern Oregon locations of Sherar's Bridge, Burnt Ranch, Mitchell, Antone, and Camp Watson. By May 1864, Henry Wheeler had established a regular stage and mail line between The Dalles and Canyon City, carrying gold dust from the mining camps for Wells, Fargo and Company. Finally, a group of investors arranged to secure a federal land grant in 1867 to subsidize road improvements over the route from The Dalles to Canyon City.[29]

The Dalles/Canyon City Military Wagon Road had its inception in the early 1860s when Congress passed legislation to build roads to support the movement of troops and supplies for the military in various states. Between 1864 and 1869, Oregon received five such grants from the public domain for road purposes. The state turned to private companies to do the construction. In the case of The Dalles/Canyon City route, Congress authorized a grant of 556,627 acres, which the state then turned over to The Dalles Military Road Company in the form of alternate sections of land, three miles on either side of the route, which extended a total distance of 357 miles. The company had five years to finish construction, and the governor of the state had to certify the road's completion before the land ownership could be transferred. In fact, the company owners did little to improve the roadway, simply grading some rough spots in the existing pathway. The governor certified the work anyway. Over time, however, settlers along the route through central and eastern Oregon questioned the legality of the governor's certification and the loss of otherwise freely available public domain for their settlement. The construction company quickly sold out to speculators who held the grant lands off the market until they could reap a big profit. The public outcry of fraud led to a protracted court battle, which the road company ultimately won in an 1893 US Supreme Court decision. Until 1910, when the land company finally put the property up for sale, it simply leased the ground to actual settlers.[30]

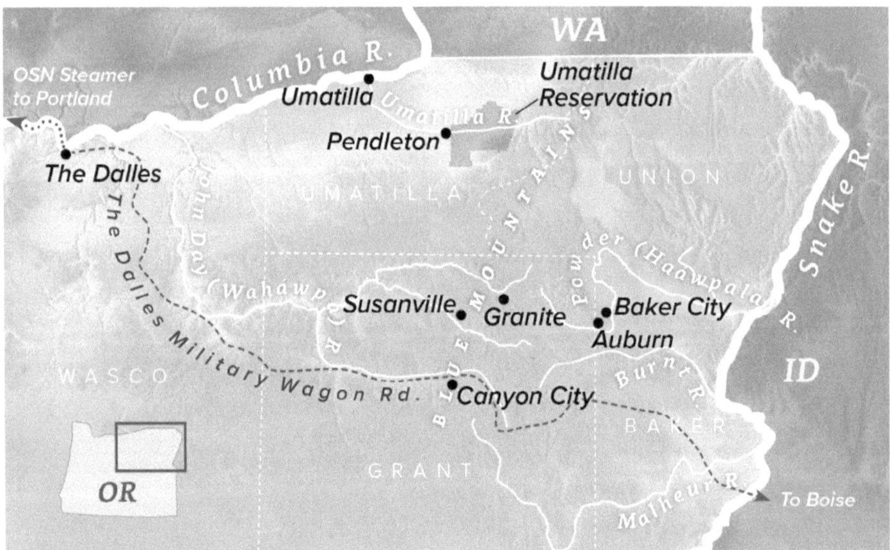

Modern map showing mining and commercial communities and the main transportation route to the mines during the 1860s and 1870s. Courtesy of the OHS and Jenny DuVander.

RELATIONS BETWEEN INDIGENOUS PEOPLES AND EURO-AMERICANS

Native Americans have lived in what became eastern Oregon for thousands of years before the arrival of Euro American miners. When non-Native people began to arrive in the early 1800s, the region was home to several tribes. By the time of the mining boom of the 1860s, most had negotiated treaties under duress, which reserved a variety of rights as well as some of their lands. The Blue Mountain region is a part of the traditional lands of the Cayuse, and the Treaty of 1855 established the Umatilla Reservation from lands ceded by the Umatilla, Walla Walla, and Cayuse tribes. Southeastern and central Oregon includes traditional hunting and gathering grounds of the Northern Paiute and Bannock tribes, whose lands extend across the Great Basin. Some Paiute bands, such as the Walpapi, resided permanently in what is now southeastern Oregon. The Northern Paiute are now confederated with the Warm Springs and Wasco, as part of the Confederated Tribes of Warm Springs Reservation, and the Bannocks now partially comprise the Shoshone-Bannock Tribes of Fort Hall. The influx of miners and settlers—yet another example of settler colonialism in Oregon's nineteenth-century development—further encroached on the Native Americans' access to their homelands and the resources they contained, leading to conflict. At the time, the Northern Paiute even considered the tribes living on the Warm Springs Reservation as enemies and periodically raided the reservation, stealing horses and cattle. The Northern Paiute responded to the new pressures on their traditional lands by attacking isolated miners, stage drivers, and ranchers during the gold rush era.[31]

In the summer of 1864, to stop the raiding by the Northern Paiutes and protect the newly arrived miners and settlers along The Dalles/Canyon City Road, the Army mounted a major offensive under Capt. John Drake of the 1st Oregon Cavalry. Drake's command consisted of 167 officers and enlisted men, 22 Native American scouts from the Warm Springs Reservation, and 39 civilian support personnel. They transported supplies by mule-drawn wagons and packtrains, utilizing 180 mules. Capt. Drake first established a string of camps and posts across the Crooked and John Day River watersheds and then conducted field operations. The key military posts for this and subsequent campaigns included Camps Maury, Watson, and Logan. Another detachment of the 1st Oregon Cavalry and infantry, under Capt. George Currey, operated simultaneously in the Owyhee and Steens Mountain country. The Army accomplished little with its 1864 campaign, as the Northern Paiutes avoided any major contact, while continuing minor raids against miners and settlers. On October 11, 1864, at the end of the field operations, Capt. Drake wrote in his journal, "The campaign is ended, the results achieved are not all that I hoped for at the beginning." Then, trying to strike a more positive note, he added, "Still, a good deal has been done. The hostile Indians have been driven out of the country near the Canyon City road far to the south. They are not likely to disturb the settlements north of the mountains for some time to come." But the Native Americans did return, and the Army under Col. George Crook again conducted relentless operations, even during the winter, against the Northern Paiutes in 1866 and 1868, with limited results. At the conclusion of operations by the 1st Oregon Cavalry in 1866, however, Adj. Gen. Cyrus Reed wrote: "Under their [the cavalry] protection, a large mining country has been developed, and millions of the precious metals brought into circulation." In 1872, some members of the Northern Paiute and Bannock Tribes agreed to go on the Malheur Reservation located in the Harney Valley of Grant County. Continued ill treatment of the Native Americans confined to the Malheur Reservation and encroachment on the reservation lands by homesteaders and cattlemen contributed to the Bannock War of 1878, which started in southwestern Idaho and expanded to eastern Oregon. Military defeat of the tribes involved in that uprising and their subsequent dispersal to other reservations led to the official closure of the Malheur Reservation in 1883.[32]

ECONOMIC IMPACT OF EASTERN OREGON GOLD

The boom in Auburn's gold production peaked around 1865. When the miners exhausted the alluvial deposits in nearby streams, their attention turned to new gold discoveries in Idaho and Montana. As people left Auburn, more permanent settlers arrived in Baker County to homestead in the Powder River Valley a few miles to the east of the mining camps, and a new town, Baker City, arose as the service center for this agricultural community. By 1868, Baker City had sufficient population and exhibited enough growth potential to take the county seat designation from declining Auburn. Canyon City fared better than Auburn. Its mining activity lasted longer, and its geographic location allowed it to remain the county seat of Grant County as homesteaders poured into the area.[33]

Some unsuccessful miners and individuals who came to provide short-term services in the mining camp soon found the valleys near Auburn and Canyon City had good agricultural possibilities. As an incentive to engage in farming or livestock raising, those still mining and numerous military personnel stationed in the region provided a ready market for foodstuffs grown locally. The passage of the Homestead Act in 1862 offered an added inducement to testing the farming potential of the region. That Act provided 160 acres of free land from the public domain to actual settlers five years after they demonstrated improvements. Experimentation proved that viable crops of grains could be grown while the nutrient-rich native bunchgrass encouraged the rise of a livestock industry.[34] State and federal census records (Table 1.2) revealed how quickly the agricultural settlement took off in the Powder River, John Day River, and Harney valleys of eastern Oregon.[35]

Table 1.2. Population of Baker and Grant Counties

	1865*	1870+	1875*	1880+
Baker County	857	2,804	1,999	4,616
Grant County	2,193	2,251	1,582	4,303

*Source: *State Census; +US Census Schedules*

The state census records also detail the growth of agriculture. Table 1.3 shows the steady expansion of the most prominent crops and types of livestock grown and raised in the two counties.

Table 1.3. Agricultural Production*

	Baker County		Grant County	
	1865	1875	1865	1875
Acres cultivated	1,155	4,492	364	6,593
Barley (bu.)	2,130	31,617	567	15,499
Oats (bu.)	7,540	58,359	8,010	31,265
Wheat (bu.)	250	10,886	384	26,605
Hay, tons	1,151	7,549	1,143	4,791
Cattle	1,153	12,341	2,783	27,597
Horses	458	2,207	524	3,879
Sheep	6	6,341	416	20,355

**Source: State Census*

The large number of cattle recorded in 1875 reflected the expansion of that agricultural activity on the public domain in the southern portions of both counties.[36]

The end of the first eastern Oregon gold boom during the late 1860s and much of the 1870s also had a detrimental effect on the economies of The Dalles and Umatilla Landing. The business of outfitting miners and shipping supplies to the mining communities of eastern Oregon and Idaho dropped precipitously. Because The Dalles served as a major center of the OSN Company's operations on the Columbia River, it experienced less of a decline than Umatilla Landing. That company's large freight warehouses and busy machine shops, which repaired and maintained steamboats and portage railroad equipment, continued to employ a large workforce. By the 1870s, moreover, the OSN Company had found new cargo to replace its lost mining supply business. Large quantities of wheat from the Walla Walla Valley in the Washington territory and Umatilla County in Oregon needed transport to

Portland for shipment overseas. In 1875, for example, the OSN steamboats transported 18,000 tons of wheat from the upper Columbia region to Portland. According to the federal census reports, Umatilla County's production of wheat grew from 28,209 bushels in 1870 to 915,571 bushels in 1880.[37]

Umatilla Landing did not fare as well as The Dalles. The loss of the eastern Oregon mining business was compounded by new competitors supplying the Idaho mines. The completion of the transcontinental railroad to San Francisco in 1869 enabled merchants in northern Utah to supply the Idaho mines more cheaply than dealers hauling merchandise overland from the Umatilla Landing. Theodor Kirchhoff, a sometimes merchant and full-time travel writer, graphically caught the decline when he passed through the town in 1872 after a nine-year absence:

> This return to Umatilla inflicts on me a peculiarly painful disappointment. For if, after roaming far and ranging wide, the traveler finds the once-bustling place moribund—what a sad surprise! . . . I despair to recall its youthful vigor while I behold its senile decay. . . . Walking to the first row of houses, I cannot believe my eyes. Freight used to be stacked everywhere, the riverfront alive with people on the move and things in transit. Now it is deserted. Umatilla I knew had fallen on evil days but I never expected such a sorry sight. Once, in lively streets, harness bells jingled while hundreds of muleteers' whips cracked. . . . But *that* Umatilla has vanished. While a lone wagon in a space of several days might take the dusty road toward the Blue Mountains and Grande Ronde valley, the Pacific railroad bustles with freight for Boise [Idaho] mines. Umatilla's artery has been cut.[38]

The federal census returns for The Dalles and Umatilla Landing captured the boom and bust of the latter and the survival of the former:

Table 1.4. Population of The Dalles and Umatilla Landing*

	1860	1870	1880
The Dalles	802	942	2,232
Umatilla Landing	1,000+ *(1863 est.)*	206	149

**Source: US Census Schedules*

How much wealth in gold and silver did Baker and Grant counties produce in their placer heydays of the 1860s? Estimates given in historical dollar values vary because the federal government did not begin to compile accurate production statistics until 1877. Most of the Oregon gold and silver output

Umatilla Landing on the Columbia River. Supply and transportation hub for eastern Oregon mines during the 1860s. Courtesy of the OHS Research Library.

reached the San Francisco branch of the US Mint from either an express company, such as Wells, Fargo and Company, or a bank, such as Ladd and Tilton located in Portland. Some gold, however, never reached the San Francisco branch; it was used as a medium of exchange, for industrial purposes or the manufacture of jewelry, or in the case of Chinese miners, kept outside the government system and sent home to China. The *Oregon Business Directory and State Gazetteer* estimated in 1873 that from 1851 to 1873, the state had produced about $40 million in gold. It noted however, that reliable figures were difficult to come by because "there is no place where they are collated." The *Directory* went on to observe:

> Many merchants in Grant County send gold dust through the mail, in order to avoid paying the heavy freight charged for expressage, which amounts to three and a half per cent, hence no knowledge of that amount can be gained. [39]

The October 15, 1864, issue of the San Francisco-based *Mining and Scientific Press* reported the gold production for Canyon City at $12 million—double that for 1863. One knowledgeable government expert stated in 1901, that Oregon's output between 1862 and 1865 might have totaled $50 million. In 1865 alone, the estimate for Oregon reached $19 million. Between 1864 and

1870, Wells Fargo reported gold and silver shipments from Portland of almost $30 million. Over the next decade, as the surface deposits of gold played out, gold and silver production dropped dramatically. As presented in Table 1.5, government experts estimated the results for Oregon and Washington between 1866 and 1876, which totaled $30.2 million. Placed in a broader perspective, Oregon mines produced roughly $60 million in gold between 1862 and 1867—the height of the placer boom—while California mines yielded $210 million during the same period. As indicated above, to translate the historical dollars into a modern value, one should multiply by the average inflation factor of about 18.11 for the decade of the 1860s.[40]

Table 1.5. Gold and Silver Production in Oregon and Washington*

1862–1865	$50 million	1871	$2.5 million
1866	$8 million	1872	$2.0 million
1867	$3 million	1873	$1.4 million
1868	$4 million	1874	$1.0 million
1869	$3 million	1875	$1.2 million
1870	$3 million	1876	$1.1 million

**Historical dollar values (Source: Brooks and Ramp)*

The Oregon state censuses taken in 1865 and 1875 provided another measure of gold production for those years. The 1865 census reported that Baker and Grant counties produced a total of 79,287 ounces of gold, which at $20.67 per ounce, would make a value of $1.6 million. By 1875, the state census listed a production for the two counties of only 4,103 ounces, having a value of $84,809. The accuracy of the state surveys is questionable, as their figures were considerably below the estimate given by the federal mining experts noted in Table 1.5.

The 1862 gold rush to eastern Oregon had long-term effects on the future growth of the region. Until that time, almost all immigrants to Oregon came with the intention of settling in the Willamette Valley to take advantage of its agricultural opportunities. With such a focus on that promised land, the Oregon Trail pioneers passed through the valleys and over the Blue Mountains of eastern Oregon, giving little thought to its agricultural possibilities. Between 1862 and 1865, thousands of prospectors and those who hoped to profit from supplying them flocked to eastern Oregon in search of great riches. Some were successful at mining, but many were not. Most of the unsuccessful strivers moved on to other Pacific Northwest gold strikes, but some stayed on, believing that they could do well developing agricultural opportunities or providing various kinds of services to those who remained. In addition,

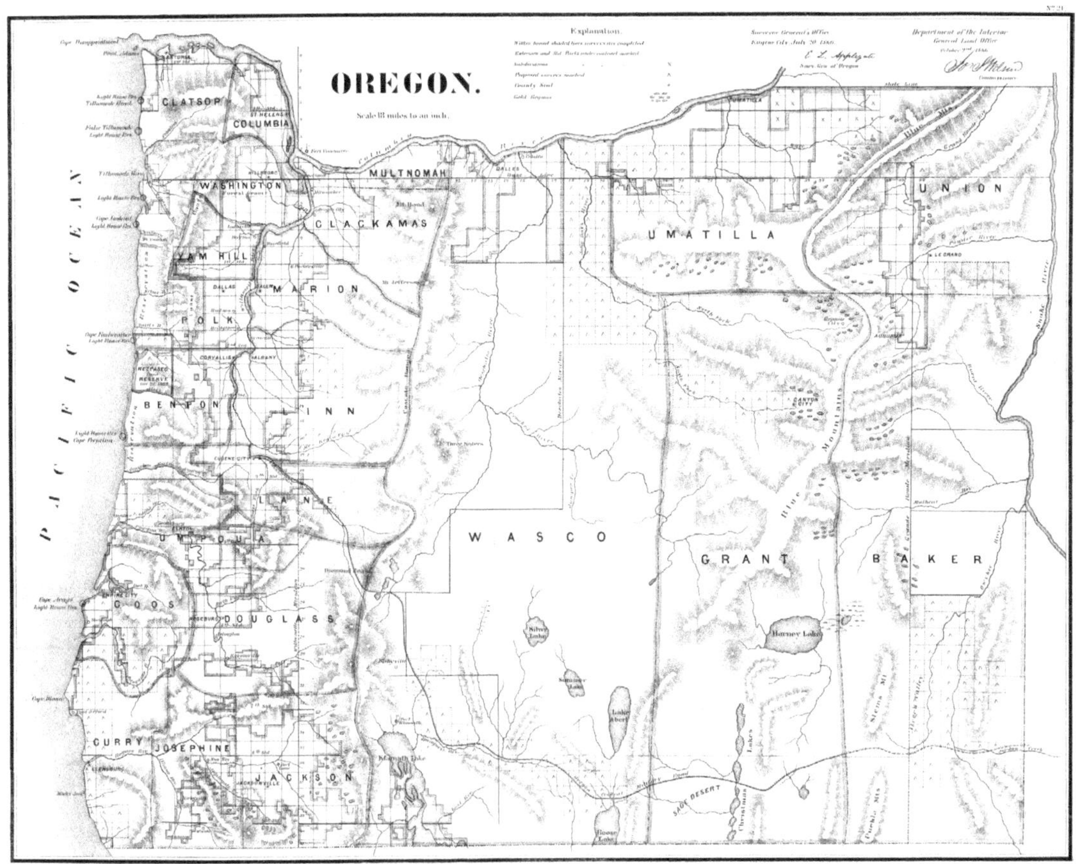

1866 General Land Office map delineating the four counties comprising eastern Oregon and locating the main mining communities. Courtesy of the OHS Research Library.

in the post-Civil War era, a new wave of settlers came west over the Oregon Trail. When these overlanders reached eastern Oregon, they learned that the best lands in the Willamette Valley and other parts of western Oregon had been taken and that they ought instead to take up claims in the unsettled valleys and range lands of eastern Oregon. Tables 1.2 and 1.3 above show that thousands acquired land and founded permanent communities in Baker and Grant counties. The two other eastern Oregon counties—Umatilla and Union—also grew quickly. Between 1865 and 1880, Umatilla increased from 1,803 to 9,607, while Union expanded from 2,334 to 6,650—both on the strength of their agricultural development, especially dry land wheat farming in Umatilla County. The livestock industry also thrived in eastern Oregon during these years.[41]

The eastern Oregon and Idaho gold industry not only stimulated the settlement of the eastern part of the state but also greatly boosted the economy of Portland. The OSN investors earned large profits, which they then invested in mining properties, farmland, city real estate, mercantile operations, transportation, and the wool and wheat trade. Portland wholesale merchants, such as Henry Failing and Henry Corbett, reaped solid profits from selling supplies to miners and small mercantile dealers in the gold camps. Some successful gold seekers returned to Portland and decided to stay permanently, investing their mining returns in local commerce and real estate. All this gold and commercial activity fueled the growth of banks. Portland's first bank, the Ladd and Tilton, increased its capitalization twelvefold over the course of the 1860s, as large deposits flowed into the bank and sizeable loans poured out. By 1870, the city had three banks.[42]

An east coast newspaper editor, after a trip to the West in 1865, succinctly captured Portland's dynamic growth based on the influx of eastern Oregon and Idaho gold:

> Portland, by far the largest town in Oregon, stands sweetly on the banks of the Willamette. . . . Ships and ocean steamers of highest class come readily hither: from it spreads out a wide navigation by steamboat of the Columbia and Its branches, below and above; here centers a large and increasing trade, not only for the Willamette Valley, but for the mining regions of eastern Oregon and Idaho, Washington Territory on the north, and parts of even of British Columbia.[43]

In 1894, the author of the *Oregonian's Handbook of the Pacific Northwest* wrote, concerning the 1860s:

> Portland, by virtue of her position, became the supply point for these diggings [in eastern Oregon and Idaho]. . . . This was a period of great business activity in which careful business men could rapidly accumulate a fortune. Thoroughly understanding the laws of supply and demand and taking advantage of the exceptional opportunities for the acquirement of wealth, the early merchants here accumulated, or were instrumental in bringing to the city, the money that has made Portland a great financial center, and in proportion to its population, one of richest cities in the world."[44]

Looking backwards from 1897, an *Oregonian* correspondent reflected similar sentiments:

> The wealthy men of Oregon, who came here 40 years ago, and amassed fortunes, had their periods of hard times and depression.

> . . . The mining industry was the real base of their prosperity. During the 20 years from 1860 to 1880, the city of Portland carried on an enormous business with the placer diggings of Eastern Oregon and Northern Idaho, which annually contributed their millions of gold dust to Portland in exchange for merchandise.[45]

As late as 1904, mining expert J. H. Fisk asserted, "It is a well-known fact that the City of Portland . . . owes its origin and prosperity to the early mining in the state, and today is being largely maintained by the more recent quartz mining."[46]

Only a few members of Portland's wealthy elite, such as John Ainsworth, R. R. Thompson, Henry Green, and Simeon Reed, invested directly in eastern Oregon and Idaho mines, but most profited indirectly through provisioning and transporting miners to the gold fields.[47] Merchants, such as Henry Failing and C. H. Lewis, personally refused to invest in mines. As Failing put it in 1865, he "thought there were entirely too many men with 'Quartz on the Brain' sponsoring wildcat mining schemes."[48] Failing reinvested his business profits to grow his Portland interests. The 1870 federal census supplied another measure of eastern Oregon gold-generated wealth during the 1860s. On the census schedules, eighteen Portland men reported total assets over $100,000, and at least five of these received their main source of wealth from the OSN and related enterprises. Most of the others operated mercantile businesses that profited from supplying the eastern Oregon gold fields.[49] Historian Arthur Throckmorton has estimated that "gold production from 1861 through 1867 provided the region an annual average surplus of some $15 million, which was immediately available for development."[50] Over time, the multiplier effect of this money financed agriculture, commerce, and transportation projects in both Portland and Oregon.

Between 1860 and 1870, Portland's population advanced from 2,874 to 8,293, an increase of 188 percent. Over the next decade, the city's numbers grew another 112 percent—from 8,293 to 17,577. While the steady flow of profits from gold mining helped to first stimulate and then sustain Oregon's growth, the harvest and export of golden grain increased it. The shipment of grain from Portland became the engine that made future economic expansion happen. Between 1860 and 1880, wheat production, a large amount of it from northeastern Oregon, grew by 800 percent. In 1868, Portland began wheat exports to Liverpool, England; and within three years, eighty-five ships focused on this business alone. In the 1870–1871 season, Portland exported wheat worth $400,000. By the 1877–1878 season, wheat shipments reached $4.6 million. To support the wheat and flour export trade, Oregon's congressional delegation prompted Congress to authorize the US Army Corps of

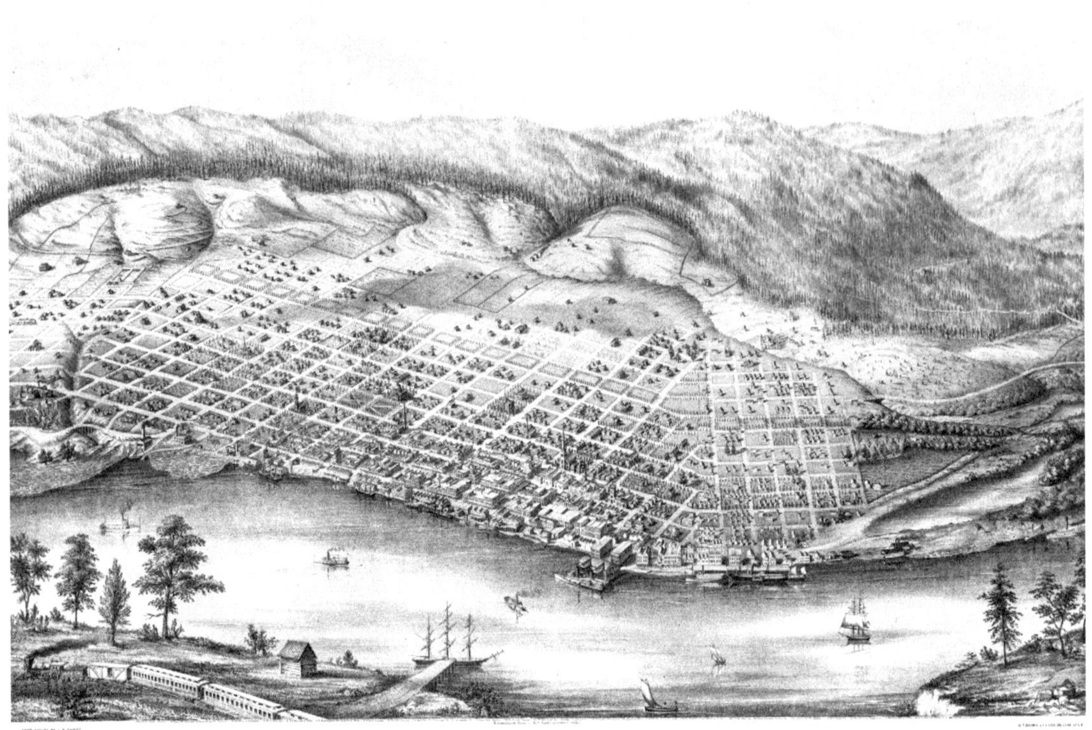

A bird's-eye lithograph of Portland in 1870, revealing astonishing growth over the previous decade because of the economic impact of gold mining. Courtesy of the OHS Research Library.

Engineers to improve navigation on the Willamette and Columbia rivers by removing rocky reefs and other obstructions and by dredging sand and silt deposits from the annual freshet. Such steady, in-river work enabled the port of Portland to accommodate increasingly larger ships. The average tonnage of grain vessels, for example, increased from 1,058 in 1882 to 1,661 by 1897. Impressively, Portland's assessed property values grew from $7 million to $18 million between 1870 and 1878, and urban capitalists built twenty major downtown buildings during those years. Gold from eastern Oregon and its reinvestment helped power the late nineteenth-century growth of Portland and Oregon as a whole.[51]

Just as placer mining sparked Portland's growth, it also jump-started the Euro American settlement of eastern Oregon. Two streams of settlers—one from California and the other from western Oregon—poured into the region in the 1860s and contended for dominance during the initial phase of development. In the early placer years, the first mining regulations adopted at Auburn and Canyon City, and subsequently at Granite and Susanville, were heavily influenced by prospectors and miners from California. The Californians, in fact, made up a sizeable portion of these new communities. The Oregonians competed with the Californians for political control. In Auburn, for example, the vote to elect the first recorder of claims came down to a contest between the California and Oregon factions, with the former winning. Humorous rivalries between the two groups enlivened the social life of the new mining communities. The presence of the Chinese, of course, added to the cultural and social mix in both California and eastern Oregon.[52]

In retrospect, the eastern Oregon mining camps of the 1860s shared social and political characteristics that set them apart from other Pacific Northwest mining centers of the period and, indeed, those gold rushes occurring in California, Australia, New Zealand, and British Columbia. In all these foreign places, the rush for gold collided with the struggle for order, and the resolution of the encounter lasted longer than in eastern Oregon.[53] The Oregon state legislature, in 1864, aided the process of community stability by officially recognizing local mining district regulations. Moreover, unlike some early mining camps in Idaho, Montana, and British Columbia, women and children arrived almost immediately at Auburn, Canyon City, Susanville, and Granite. The Oregon camps at Auburn and Canyon City started schools in the fall of 1862—just months after the first mining commenced. Granite and Susanville began schools in 1863 and 1864, respectively. Additionally, because Oregon had already attained statehood, the eastern Oregon mining communities achieved local government and social stability through the establishment of counties more quickly than was the case in the mining camps of Idaho and Montana. There, the process of gaining territorial status made such institutional arrangements more difficult to establish. Economically, as indicated above (see Table 1.3), agriculture and commerce in the Powder and John Day River valleys got a quick impetus from a mining population dependent on foodstuffs and other necessities provided by others. Subsequent settlers would build on this economic foundation. The mining of gold had a powerful effect on Oregon's urban and rural population growth and its economic and institutional development in the 1860s. After 1885, eastern Oregon lode gold mining would add further stimulus to Oregon's economic advancement.

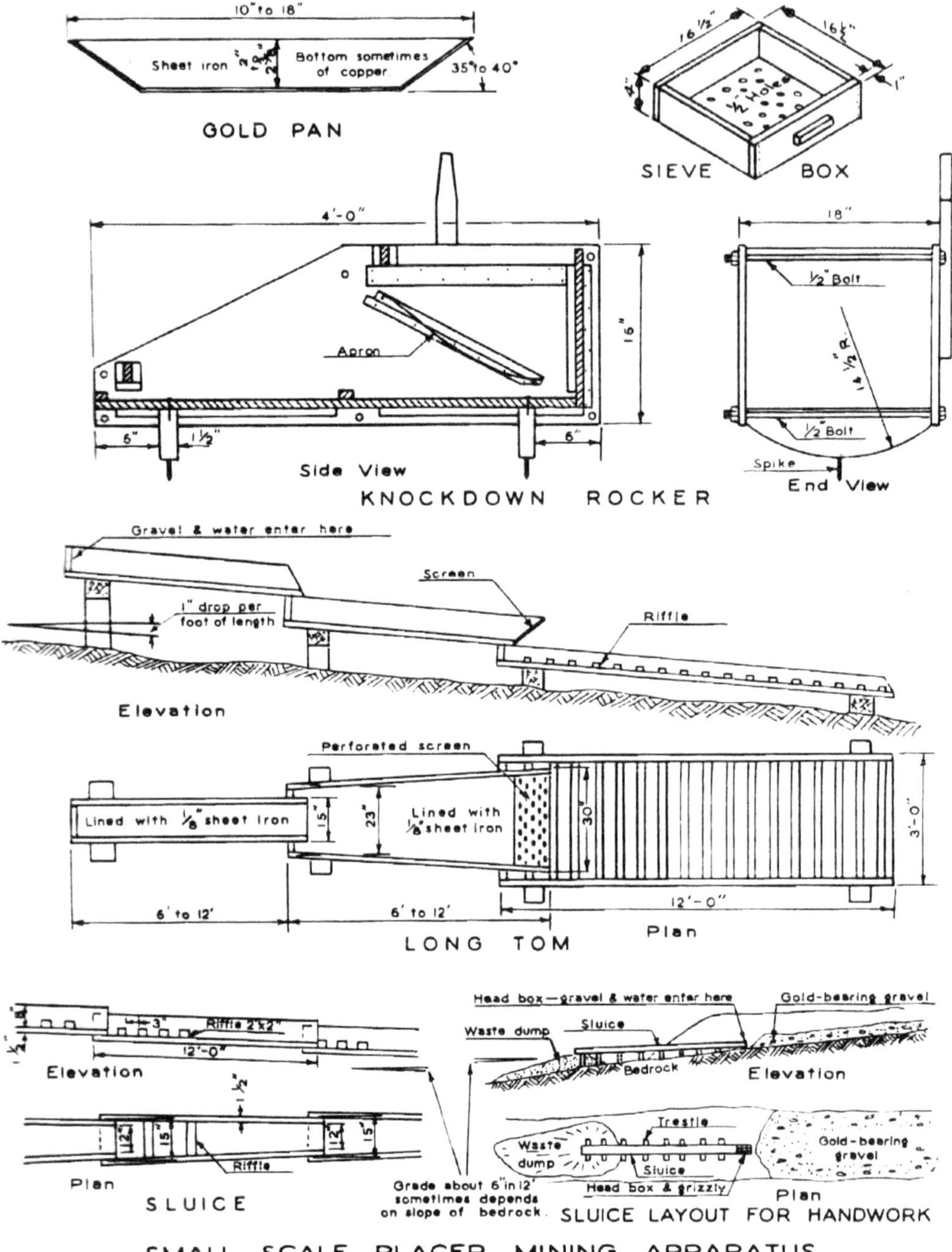

Examples of various apparatus used in placer mining. Courtesy of the Oregon Department of Geology and Mineral Industries.

Chapter Two
The Enterprise of Gold Mining

Eastern Oregon precious metal mining resulted from the interplay of both natural forces and human ingenuity. Enormous geological processes over eons of time created deposits of gold and silver, but it only took fifty years of systematic human exertions to retrieve the most valuable accumulations. Once prospectors located the surface gold and silver, primarily found in or near streambeds, miners used simple methods to recover it. Locating and removing the deeper sources of gold and silver that miners found underground required greater technical knowledge and more sophisticated methods to retrieve and process them. Precious metal mining ushered in eastern Oregon's exposure to the modern industrial age and global markets.

GEOLOGY OF EASTERN OREGON

The Blue Mountains, which extend approximately 50 miles in width and 120 miles in length, occupy about 15,000 square miles, slightly less than a sixth of Oregon's land mass. Rising from 1,000 to 9,000 feet, the mountains have a composition consisting of marine sedimentary, volcanic, and intrusive rocks. The oldest sedimentary rocks date from 300 million to 150 million years ago (late Jurassic period) and were formed on the floor of an ancient ocean. Through plate tectonics, these rocks were pushed to become part of the North American continent. The movement of these deeply buried rocks caused intense folding and faulting, which changed their original mineralogy and structure to become greenschist facies.[1]

The next major rock bodies that appeared in the developing Blue Mountains consisted of granitic material. Geologists call the largest granitic blocks batholiths, while smaller ones were referred to as stocks. The granitic rocks formed from magma that invaded the older metamorphic rocks 130 to 120 million years ago (late Jurassic to early Cretaceous period). Most of these granitic bodies were quartz diorite or granodiorite.

The succeeding major rocks were labelled tertiary and consist of lava flows, tuffaceous rocks, and sedimentary deposits. The lava and tuff appeared as basalt or rhyolite. The sedimentary deposits formed in freshwater lakes and river floodplains. The tertiary rocks covered the ancient metamorphic and granitic rocks.

During the Cretaceous and early Tertiary time (150 million to 30 million years ago), the area that is now eastern Oregon underwent uplift, volcanic eruptions, and heavy erosion. Gold deposits or veins occurred near the edges of the granitic intrusive rocks associated with fracture zones along fault lines caused by uplifts. The fracture zones created spaces that allowed for the intrusion of magma carrying metals and other ore minerals to penetrate and, as the fluids cooled, depositing the dissolved material to form a vein. The veins could be straight or irregular and extend to great depths. Subsequent geological action could deform them. Since the igneous process took place at considerable depths, exposure of the deposits at or near the surface occurred only after long periods of uplift and erosion.

A correspondent to the *Mining and Scientific Press* in 1904 applied the above generalized account of geological processes to the specific creation of the complex gold deposits dispersed throughout the mountains of eastern Oregon:

> No important gold or silver deposits are found in the recent lavas. They are confined solely to the older rocks. Moreover, all veins containing gold or silver appear to be massed near the contacts of the sedimentary rocks with the granular intrusives. They do not necessarily follow the line of contact, but are found indifferently in either the sedimentaries or intrusives, continually branching and crossing the contact, the fissures dying out at one place only to be replaced by other parallel fissures. They are usually found as a number of local aggregations of rather short, irregular veins, and not as a well-defined and continuous belt. It is these separate occurrences that constitute the [various mining] districts. Each district may consist of several mines, usually similar in formation.[2]

TYPES OF MINES AND EXTRACTION PROCESSES

The gold deposits existed as either lode or placer in nature. The gold in lode deposits rested in solid rock that had to be dug out of the ground and processed to recover the gold. Placer gold occurred as loose flakes, fine particles, or nuggets that had eroded from solid rock and accumulated in surface sand, silt, and weathered rock, chiefly gravel, in streambeds and streambanks. The richest pay dirt often lay next to bedrock, where the solid layer held the gold dust that had filtered the deepest. An 1870 report by US Commissioner of Mining Statistics, Rossiter Raymond, succinctly captured the natural processes that produced placer gold:

> Frost, ice, and mountain torrents, aided by the decay of rocks, have broken down the veins and liberated the gold, leaving it distributed under the gravel and sand in the beds of ancient and existing streams. The force required for breaking up the rocks and veins has been expended and the work of the placer miner is rather to clean up or harvest what nature has already mined for him.[3]

To retrieve gold from placer deposits, the miner relied on running water and the fact that gold was heavier than the surrounding debris in which it rested. The simplest tools for working a placer deposit were picks and shovels and the miner's pan: a round, shallow metal container about eight inches deep and twelve inches or more across the bottom. After excavating the dirt and gravel, the miner used water to wash away the debris with a swirling motion of the mixture, leaving the much heavier gold behind at the bottom of the pan. If the panning process indicated the potential for a substantial deposit of gold, the next step required greater effort. Miners then built wooden devices such as a sluice box, a slightly inclined ten- to twelve-foot-long and one-foot-deep wooden trough with transverse ridges (called riffles or cleats) across the bottom to collect the heavier gold pieces left after the lighter dirt had washed away by a current of water from a ditch. Depending on ground conditions, the amount of water available, and the number of miners at work, sluice boxes could be strung together for hundreds of yards. For small operations, miners also used wooden rockers or cradles open at each end, which were sluice boxes with a detachable fabric sieve for screening out coarse material ahead of the metal ridges in the box where the gold settled. Even this simple device worked best with two men: one to dig and carry dirt, while the other poured water from a stream and rocked the box. In the early 1850s, California miners began developing a hydraulic process, in the form of a powerful stream of water from a hose fitted with a metal nozzle, to wash away high banks of gravel and feed the sluice boxes. The key to a successful hydraulic operation required conveying a sufficient quantity of water via ditches and wooden flumes and then sending it from a height of 100 feet to 200 feet through pipes to the hose.

As a commentator wrote in the *West Shore*, "None of the [placer] mining methods, . . . were the product of one man's invention, but have been gradually evolved from the requirements and experiences of miners in every gold field on the Pacific coast."[4] Unfortunately, poor planning and sloppy work in placer mining left waste piles containing significant amounts of gold, but the gold was hard to extract from the detritus. Moreover, other valuable deposits could lie under waste piles and thus be overlooked.[5]

Government mining experts in the 1860s criticized the early placer miners for their wasteful practices, which they argued resulted in the loss of potential national wealth. Two reasons accounted for this problem. First, most prospectors were in the mines to make as much money as possible in the shortest time before returning home to other employment. Secondly, the heavy transportation and living expenses necessary to open a new and remote mining field meant only the richest gravel could be worked profitably, resulting in the enormous wastage of lower-grade ground. Federal mining investigator J. Ross Browne went so far as to write in an 1868 commentary on placer mining, "The question arises whether it is not the duty of government to prevent, as far as may be consistent with individual rights, this waste of a common heritage, in which not only ourselves but our posterity are interested."[6] The best returns from placer mining occurred in California between 1849 and the 1860s, with hydraulic mining dominating between the 1860s and 1880s; in the other parts of the West, alluvial mining had its most productive times from the 1860s to 1890s.[7] By one estimate, the entire West produced about $1.6 billion in placer gold between 1848 and 1931, with most ($1.175 billion) coming from California.[8]

In the early twentieth century, placer miners introduced dredging, a more efficient and cost-effective approach to working large acreages of placer gravels. The most successful type of dredge, the bucketline, consisted of a wood or steel-hulled barge that contained a structure mounting a continuous chain of steel buckets for excavating, a screening and washing plant for separating the gold from the gravel, and conveyor belts for discharging and stacking tailings behind the boat. Brooks Hawley, author of a book on the Sumpter Valley (Oregon) dredges, supplied a succinct definition of a gold dredge: "It is a mechanized placer mine to get at gold-bearing gravel beneath the river level." He further elaborated, "A dredge is a boat floating in a pond of water that it automatically makes as it devours land in front of it to turn it into mud to wash downstream and to dump the washed gravel behind it." As the dredge's front buckets dug new rows in the earth, river water flowed into the work area, keeping the dredge afloat and moving forward. Finally, according to Hawley, dredging is distinguished from lode mining in that "the dredge is not a mill to crush rock to recover gold from ore but it is a washing plant to recover loose

A scene of hydraulic mining on Elk Creek, Oregon. Jets of water from the iron nozzles blast the riverbanks. Courtesy of the Grant County Historical Museum.

gold from among the gravel, gold that originally was farther back in the hills in prehistoric river channels or within solid rock before the action of the ages wore down the hills and the rocks." Although expensive to build, the final Sumpter dredge cost only a few cents per yard to operate. It excavated to an average depth of 18 feet and employed between twenty and thirty men.[9]

To recover gold from a lode deposit, miners had to drill and blast underground shafts and tunnels (adits), dig out the gold-bearing rock, and then bring it to the surface in carts. Once on the surface, the miners processed the ore by crushing, grinding, and milling it and then shipped it to a smelter to recover the gold. Most quartz veins were too small or lacked enough gold per ton to sustain a profitable mining operation. Productive veins varied from a few inches to tens of feet in width and from hundreds of feet in length to 4.5 miles in the case of the North Pole/Columbia lode located in Baker County. The greatest mine depth in the Blue Mountains reached 2,500 feet below the outcrop, again at the North Pole/Columbia lode. The gold content of most ore averaged one-half to one ounce per ton. Veins rarely exhibited uniformity in value. Most ore bodies were interrupted at irregular intervals by zones with little gold. The parts of the veins holding mineable quantities were called "ore shoots" or "ore bodies." The direction of the vein as measured on a horizontal surface was referred to as the "strike," while the angle the vein inclined below the surface was its "dip."[10]

Once recovered, miners processed the ore using a variety of equipment depending on whether they encountered free-milling or refractory gold. The former yielded to amalgamation methods, while the latter required more complicated chemical treatments to release the gold from its matrix. The earliest, simplest method used an arrastra, a wooden device that ground ores by dragging a heavy stone around a circular bed of stones. A mule or horse usually powered the mechanism, which consisted of a vertical spindle with arms attached by a chain or cable to heavy stones crushing the ore.

Another common method of working the ore involved a stamp mill. Workers first prepared the ore by breaking it into chunks the size of apples and then feeding it into the mill. The mill then further crushed the ore into a fine powder by striking it with rows of heavy iron pestles (stamps), working mechanically in a large iron mortar resting on a concrete block. Workers using water or steam driven machinery raised and dropped the stamps—weighing several hundred pounds each—six to eight inches deep, sixty to a hundred times a minute. The next step involved spreading the pulverized ore on copper plates treated with mercury, which combined with the free gold to form an amalgam. The gold was later recovered by distilling the mercury, some saved to be used again. The demand for mercury in mining was enormous, and producing it became an industry itself. By 1890, mercury output in the United States reached 120 million pounds.[11]

Yet another milling process used flotation. This method required mixing finely ground ore with water that had been treated with certain chemicals. The workers fed the liquid into tub-like cells and agitated the mixture to create a froth. The froth floated the valuable minerals to the top of the cell, while waste sank to the bottom, where it was drawn off. An alternative ore treatment procedure mixed the finely crushed ores in vats with a solution of sodium cyanide. The gold dissolved to form sodium gold cyanide, and that solution was treated with zinc or aluminum, which caused the gold to precipitate for recovery. The various chemical processes had the disadvantage of leaving behind dangerously polluted tailings, ground, or water bodies. The initial experimentation and perfection of the chemical processes used to recover refractory gold occurred during the 1870s and 1880s. The cyanide process was introduced in American mines in the early 1890s.

Mining engineers and industrialists devised new equipment and technologies to meet the challenges of gold and silver lode mining in the last one-third of the nineteenth century. For example, the invention of the Burleigh Drill

Vanner Tables at the Red Boy Mine, used in ore processing to extract the gold from its rock matrix. Courtesy of the Baker County Library.

in 1870 (and later the Ingersoll and Rand drills) revolutionized the process of rock removal. The *Blue Mountain Eagle* reported in March 1901 that the Golconda Mine had installed three Ingersoll-Seargent drills and planned to add more. A few months later, the newspaper noted that the Concord Mine had ordered several such drills and compressor plants. Operated by compressed air or steam, these mechanized drills replaced the laborious, time-consuming hand drilling required to make holes in the surface rock so that blasting powder could be inserted and fired. The invention of dynamite in 1866 and the safety fuse vastly improved the initial retrieval of ore-bearing rock, replacing ordinary black powder. Other advancements in underground mining included the development of square-set timbering to allow work amid unstable rock, the use of power hoists and wire rope to take workers underground and bring rock to the surface, and the application of high-powered pumps for removing water from the mine tunnels. The installation of square-set timber supports for tunnels was also a major part of mining labor. Workers made each set, which formed an arch, from three timbers, 8 feet long and 8 inches square, and placed the sets about every 5 feet, with braces between each for increased stability. The new mine technologies increased production and profit, but they did not necessarily make the work underground safer.

A miner operating a power drill at an eastern Oregon mine. Courtesy of the Baker County Library.

Finally, in the early 1900s, hydroelectricity began to replace steam (produced by wood-fired boilers) or waterpower as the energy source to operate the mine machinery.[12]

The adoption of electricity helped to lessen the environmental impact of lode mining on the surrounding forests. Previously, wood-fired boilers used prodigious amounts of timber. The Cornucopia Mine in Baker County, for example, needed 800 cords of wood a year in its early period of operation. Since a cord of wood represented a volume four-by-four-by-eight feet of firewood, hundreds of eastern Oregon mines consumed astounding amounts of wood. Until the federal government established the national forest system in the early twentieth century, the mines could freely cut timber for their operations on the adjacent public lands.[13]

When placer mining transitioned to lode operations, it led to a change in the type of miner and the means of accomplishing the new, more complex mining methods. The first miners on the ground at a placer location staked their claims and usually worked independently or in loose partnerships. As the initial miners in a camp rapidly skimmed the easily reached placer gold and then frantically moved on in search of the next big strike, officials began relaxing the prohibition against Chinese miners. The Chinese soon entered and, over time, profitably reworked the placer ground that they leased or bought from white miners or mined new areas they prospected independently. The methods used by the Chinese placer miners are discussed in chapter six. Next, the transition to quartz or lode mining led to a key shift in labor and management personnel. The change involved the displacement of the lone prospector or miner, who, lacking the financial resources to pursue veins underground, gave way to a more complex labor and management structure. Under the new mining conditions, corporate organizations emerged. Mine owners found that the heavy equipment and production procedures needed for quartz extraction, the cost of transporting the needed equipment, and the necessary engineering knowledge required large amounts of capital and corporate methods that only stock companies could provide. The workers for the new company-owned lode mines came from among the ranks of hired or wage laborers. The various jobs required in lode mines included muckers, drillers, powder men, timbermen, carpenters, millmen, blacksmiths, hoist operators, water boys, foremen, and office clerks. Wages, ranging between $2.50 and $9.00 a day, were based on the level of skill required. The timing and effects of the various financial, technical, and work-related changes in eastern Oregon mining will be discussed in subsequent chapters.[14]

Surprisingly, the eastern Oregon press provided little coverage of working conditions in the region's lode mines. Mining historians, however, have shown that such underground labor throughout the metal mining West was dirty, difficult, and dangerous. Working ten-hour days deep below the surface and surrounded by darkness, foul air, and often wet, slippery surfaces was challenging and sometimes unsafe. Initially, miners employed mostly hand tools such as picks and shovels, chisels and wedges, and sledgehammers to get out the ore. The use of dynamite and the introduction of mechanical drills and other powered equipment increased the possibilities for an unsafe work environment. Accidents or deaths resulted from cave-ins, poisonous blasting gases, rock dust, fires, intense heat, unstable footings, and inadequate timbering in the shafts. Unfortunately, workers suffering accidents had little recourse in courts against employers because the common law of liability (assumed risks, contributory negligence, and the fellow-servant rule) shielded mining companies from damage suits. States, moreover, were slow to adopt mining safety laws or regulations. The implementation of the eight-hour workday for miners, for example, came mostly during the first decade of the twentieth century. Miners, especially in Colorado, Nevada, and Idaho, did attempt unionization and strikes to improve their working conditions but appear to have made little use of such organizations or actions in eastern Oregon. The *Blue Mountain Eagle* made passing reference in October 1901 to a union being organized by miners at Susanville. Between 1899 and 1906, the *Sumpter Miner* reported

Fraser & Chalmers

Builders of the

GENUINE FRUE VANNER

Used by the Eureka Hill Co. after testing all different makes of Concentrators.

THE EUREKA HILL CO.'S MILL, EUREKA, UTAH. BUILT BY FRASER & CHALMERS.

Among other big mills using GENUINE FRUE VANNERS are

The Alaska Treadwell-96 Frue Vanners. The Creston Colorado-52 Frue Vanners
The Alaska Mexican-48 Frue Vanners. The Montana Mining Co-48 Frue Vanners

There are thousands more, but the above list may be enough to convince mine managers that their interests will be best served by avoiding experiments with imitations and inferior designs, and buying the GENUINE FRUE VANNER OF

FRASER & CHALMERS CHICAGO AND LONDON

A diagram of a typical gold mill operation showing the equipment used to extract gold from its ore matrix. Courtesy of the Baker County Library.

on only two strikes, one of which occurred at the Cracker Creek mines over housing conditions and the other in the Greenhorn District involving wages.[15]

THE RISE OF THE MINING ENGINEER

The increase in lode mining throughout the American West in the 1870s led to the demand for trained engineers to oversee the evaluation, development, and operation of those underground mines. At first, these engineers received their education at schools in France and Germany: the Ecole Imperiale des Mines in Paris and the Konigliche Sachsische Bergakademie in Freiberg. In the United States, mining education took off with the implementation of the 1862 Morrill Land Grant Act, which supported state agricultural and mechanical colleges. Still, foreign mining engineers worked extensively in the American West, while United States mining schools gradually ramped up the training of technical experts. At least twenty American schools offered courses in mining, and between 1867 and 1892, they graduated 871 mining engineers. From 1870 to 1920, moreover, the number of trained US engineers in all specialties grew from 7,000 to 136,000. Many of these engineers worked overseas as well as in the United States, and knowledge exchange became transnational.[16]

The advent of national and international professional organizations accompanied and furthered the rise and ambitions of mining engineers. As one historian has succinctly put it, "Academic, technical, and professional social networks emerged to share knowledge, solve problems, and systematize the application of new technologies to the extraction of mineral resource."[17] In the late nineteenth century, more than twenty institutes for mining arose internationally, with members mobile and keenly interested in sharing technology and practice. In the United States' post-Civil War period, professionals established forty-eight scientific societies; before 1860, only fifteen had been formed. While structural, hydraulic, and sanitary engineers found a home in the American Society of Civil Engineers, a subgroup of mining engineers organized the American Institute of Mining Engineers in 1871. This organization embraced a broad membership of professionals and nonprofessionals interested in all phases of mining. By 1905, it had a membership of 3,600. The expanding membership was fed by the ever-growing development of new goldfields worldwide. For example, between 1848 and 1891, gold mines across the globe produced 435 million ounces of gold.

Just as the new engineering societies and institutes helped facilitate the spread of mining technical knowledge and expertise, so did the medium of technical journals and magazines, which grew in numbers and circulation after 1860. In the United States and abroad, professional organizations published annual transactions and other scientific publications. The professional technical publications were joined by weekly newspapers appealing to a broad range of readers interested in mining information. The most important of the mass media publications focusing on mining news included the *Engineering and Mining Journal* out of New York and the *Mining and Scientific Press* published in San Francisco. The *Engineering and Mining Journal* appeared in both New York and London editions beginning in the 1890s. These subscription-based publications provided a compendium of current practice, new machinery and refining processes, geological reports, and international travelogs designed to appeal to professional engineers, practical (but often untrained) miners, and investors. In 1891, with not a little national hubris, the *Engineering and Mining Journal* announced its mission as "the creation of a technical literature that will make the whole world recognize the preeminence of American practice."[18]

The rise of American mining engineers directly influenced national mining policy as well as the practice and business of mining itself. In the immediate post-Civil War period, the federal government, and especially Congress, took a new interest in furthering the exploitation of the nation's mineral

Rossiter W. Raymond (1840–1918), a leading American mining engineer, editor, and author. Courtesy of the OHS Research Library.

wealth in the West. While Congress enacted the first federal mining laws, the Department of the Treasury, to gain a better understanding of current mining practices and the potential of a major economic resource, appointed Rossiter Raymond as the commissioner of mining statistics to annually survey and report on the field. Raymond, after graduating from the Frieberg mining school, returned to the United States and established himself as a leading consulting mining engineer. He applied his training and experience in his new position as commissioner of mining statistics to gather and publish technical information and provide field data that could encourage the mining industry to be more productive and less wasteful. His reports, between 1869 and 1876, represented the most thorough studies available of the geology and conditions of mining districts in the mountain West. He personally visited the gold fields of eastern Oregon in 1869. In his early reports, Raymond pointed out that placer miners too often carried out their work "in a lawless and careless way, without regard for the future." He believed that profitable mining in the West lay in the scientific development of lode mines: "When the industry of mining in these rich fields is based upon a foundation of universal law, and shaped by the hand of educated skill, we may expect it to become a stately and enduring edifice, not a mere tent, pitched to-day and folded to-morrow." This would entail learning how to go deep underground and how to properly reduce lower-grade ore to extract the gold. In subsequent reports, he provided policy recommendations, descriptions of advances in mining knowledge and technology, and promoted the establishment of a national mining school.[19]

FEDERAL MINING LAW

Both natural and man-made conditions influenced the course of mining throughout the West. Miners had no control over the geological processes that created gold, but they could and did establish the regulations and laws governing the location and extraction of that precious metal. After discovering gold-bearing land, the prospectors typically met in an open body and drew up the rules for locating and working a claim and resolving potential conflicts that might develop. Before 1866, no federal law governing mining operations existed. The prevailing approach to the problem of protecting property rights in mining claims and their development followed the experience and procedures worked out by miners during the California gold rush of the 1850s and the application of traditional common law principles of property. By the mid-1860s, western corporate mining concerns began calling for the federal government to bring greater clarity and legal protection to mining rights.

In 1866, Senators from California and Nevada and a territorial delegate from Colorado—representing mining interests in their states and territory—got Congress to pass legislation establishing rules for lode mining claims. The legislation declared that the public domain was open for prospecting and mining by all citizens and aliens who had declared an intent to naturalize. The law, in effect, recognized the mining claim as a property right free of royalty. It also incorporated all existing state, territorial, and local mining district regulations without change into the new law. Most importantly, the measure adopted the apex rule, which held that a claimant possessing the apex (high point) of a vein had the right to pursue the vein outside the sidelines of the claim. The statute limited the claim size to 300 feet along the vein and allowed the claim to be patented (taken into private ownership) for $5.00 an acre. The limit on claim size caused problems, however, when owners made competing claims on the same underground ore lodes. Also, the law did not require a specific amount of ongoing work to maintain the title to the claim. In 1870, Congress enacted another mining law, this time addressing placer mines. This legislation allowed placer claims of twenty acres and associations of up to eight individuals to claim 160 acres. A claim could be patented for $2.50 per acre.[20]

In 1872, Congress took another look at mining rights and enacted a statute that provided for a uniform system of location, recording, and working mining claims. The law declared public lands open to mining, either placer or vein, and lode claims could be located as part of a placer claim. The claimant could live on a claim and develop an area up to 1,500 feet long and 600 feet wide, centered along the middle of the vein. The end lines of a claim had to be parallel, and where lode claims intersected, the senior one prevailed. The apex rule still applied (the right to follow a vein outside the claim boundaries under certain circumstances), but a locator did not have the right to enter the surface of a claim owned by another. Extensive faulting made tracing a vein continuously underground from its surface apex challenging and led to many lawsuits. Tunnels excavated for the development of a lode conferred on the tunnel owners the right to all veins or lodes within 3,000 feet from the face of the tunnel. Claimants had to post a legal notice on the ground with distinctly marked boundaries. The law allowed miners up to five acres for milling purposes, which could be located on non-mineral lands. To keep the legal right to a claim, the miner had to spend $100 annually working it. If a claimant failed to perform the annual work requirement, the claim was considered abandoned and could be relocated by others. Patents for lode claims cost $5.00 an acre and half that amount for placers, plus any filing, surveying, recording, and publication costs. Finally, the statute specified that local law applied to claims taken up prior to 1872, and all claims filed after that date had to follow the new law. This comprehensive statute remains the law pertaining to hard rock mining to the present day. Over time, much litigation would occur to interpret the law's meaning in practice.

The senators promoting the Mining Law of 1872 argued that it merely applied the concept of the Jeffersonian "noble yeoman farmer" to the miner. Both were making America productive and its citizens economically independent by developing nature's gift of abundant natural resources. As historian Gordon Bakken succinctly put it: "The mining law was continuity, democracy, entrepreneurial opportunity, and wealth producing without burdensome taxation imposed on the poor miner."[21] The Mining Law of 1872, however, had many unintended legal, economic, and environmental consequences that would play out over the next 150 years in the American West.

THE HUNT FOR GOLD

The search for gold was a risky and complex undertaking, especially for lode claims. A column in the *Mining and Scientific Press* outlined the basic and often daunting steps involved:

> Vein mining is a very complicated business; one in which it is easier to throw a away money, perhaps, than in any other immense excavations in solid rock and heavy timbering, requiring judgment

Columbus Sewell, a Black man who was a teamster and prospector living in Canyon City in the late 19th century.

> and skill—not always at command—expensive machinery to raise and crush or smelt the ore, pump out the water and raise and lower the men, and finally, thorough acquaintance with the science of chemistry and metallurgy in order to separate the gold from the ores, are requisite to success. And all this, even when the vein is rich, well developed, even and continuous. Very few metalliferous veins are equally rich for any considerable distance, either lengthwise or up and down; neither are surface indications always to be trusted. The consequence of all which is a double quadruple uncertainty. . . . The natural and unavoidable risks encountered, are quite sufficient to deter all but the most venturesome from risking their capital in such undertakings.[22]

To fully comprehend the process of finding and recovering precious metals, it is first necessary to recognize the difference between a prospector and a miner. Simply put, the prospector searched for gold while the miner worked at recovering what the prospector had identified as possible riches. The annual quest for gold usually began in the springtime. Once the snow was off, prospectors began their search in either existing areas known to have gold or in new regions suspected of having such wealth. If they found enough evidence of paying ground, they could stake a claim and carry out the initial assessment work necessary to hold it. For most prospectors, however, it was the thrill of the hunt and discovery that excited them, and they did not care to do the sustained work necessary to prove the full extent or value of their find. Since a mere claim required development to ascertain its true worth, the prospector seldom made much money when he sold it. Then, he was on to the next tempting location, for the prospector believed that if he found gold once, he could do it again. As a correspondent wrote in the *Mining and Scientific Press*:

> While he [the prospector] swears by the claim he happens to have in hand at the time, and thinks it promises better than any he ever had, he also thinks some other country better than the one he is in. The influence of habit makes the old prospector restive. A few times over the same ground gives it a played-out look to him. He sighs for new regions, new surroundings and a change of base. He eagerly listens to all that is told about newly discovered regions and swallows stories that sober reasoning and thought tell him cannot be true. . . . The fascination of a new camp is one that the prospector finds it hard to resist; and when half a dozen men leave a camp for a new one, half of those left begin to think about it too.

As the correspondent went on to lament:

> Of course, no amount of advice would accomplish much. Each man knows what he wants better than any other man. Prospectors will keep on roaming and changing till time is no more.[23]

The miner or mining investor, on the other hand, usually paid only a small amount for a claim, since the true value was yet to be determined. He then systematically worked the placer ground or sunk the shafts and dug the tunnels necessary to follow a vein to determine its extent underground. If a claim turned out to be valuable, the miner owner had to recover the ore, mill and refine it to separate the gold, and then ship the precious metal to the mint for sale—altogether an expensive and laborious process. What every miner soon needed to make the most of his claim was more capital and equipment. Some miners tried to raise the necessary capital on their own or sold out to others who had or could gather the funds needed to develop a paying mine. By the end of the period under study, the lone prospector and small mine operator had largely been displaced by the corporate mine operator. The rise of professional mining engineers provided the advancing technical knowledge of lode mining, milling, and ore reduction that helped to accomplish the transition.

Finally, whether one was a placer or a lode miner, there was a certain irony in the material conditions of the ore reserves in which they worked. As mining historian Kent Curtis explained:

> Mining men and partnerships organized capital to coordinate the extraction and processing of valuable ores, but the more ore that was removed and stamped, the less ore remained in the ground to be mined. For placer gold miners, this fact led them to work over the region of their claim and then move on. For lode miners, it meant they had to push their mines deeper and deeper into the earth to access more of the ore lode, if it existed at all.[24]

The push underground would mean encountering new types of conditions, requiring innovative kinds of knowledge and skills to address them.

Famous Monumental Silver Mine near Granite, Oregon. One of the oldest developed lode mines in Grant County. The mill was built in 1878–1879. Courtesy of the Baker County Library.

Chapter Three
Placer Mining Camps at Granite and Susanville

Prospectors heading to the Idaho gold fields and those fanning out from Auburn and Canyon City soon explored every river, stream, canyon, and hillside in the Blue Mountains. In the process, they located rich strikes at Granite Creek, a tributary of the North Fork of the John Day River, in 1862 and Elk Creek (Susanville), a tributary of the Middle Fork of the John Day River, in the spring of 1864. Thus began a thirty-year period of intense placer mining in eastern Oregon. Prospecting for the ledges and underground sources of the alluvial gold also continued during the 1870s and 1880s. Lode mining finally took off in the late 1880s, after the arrival of the transcontinental railroad made it possible to bring in the heavy equipment needed for underground mining.

After describing the topographical setting of mining and the previous use of that terrain in eastern Oregon by Native Americans, this chapter will focus on the persistent search for and recovery of placer gold and the gradual transition to lode development, especially at Granite and Susanville. The discussion reveals the limits that transportation and the lack of sufficient capital investment placed on mining expansion and profits. A deep dive into the census records for Granite and Susanville will show the contours of the ethnicity, gender, age, and family structure of typical eastern Oregon mining communities prior to 1890.

A MOUNTAINOUS SETTING

When entering the section of the Blue Mountains containing Granite Creek and Elk Creek, the first prospectors encountered a true wilderness of rugged mountains, thickly timbered forests, and myriads of deep drainages. At the

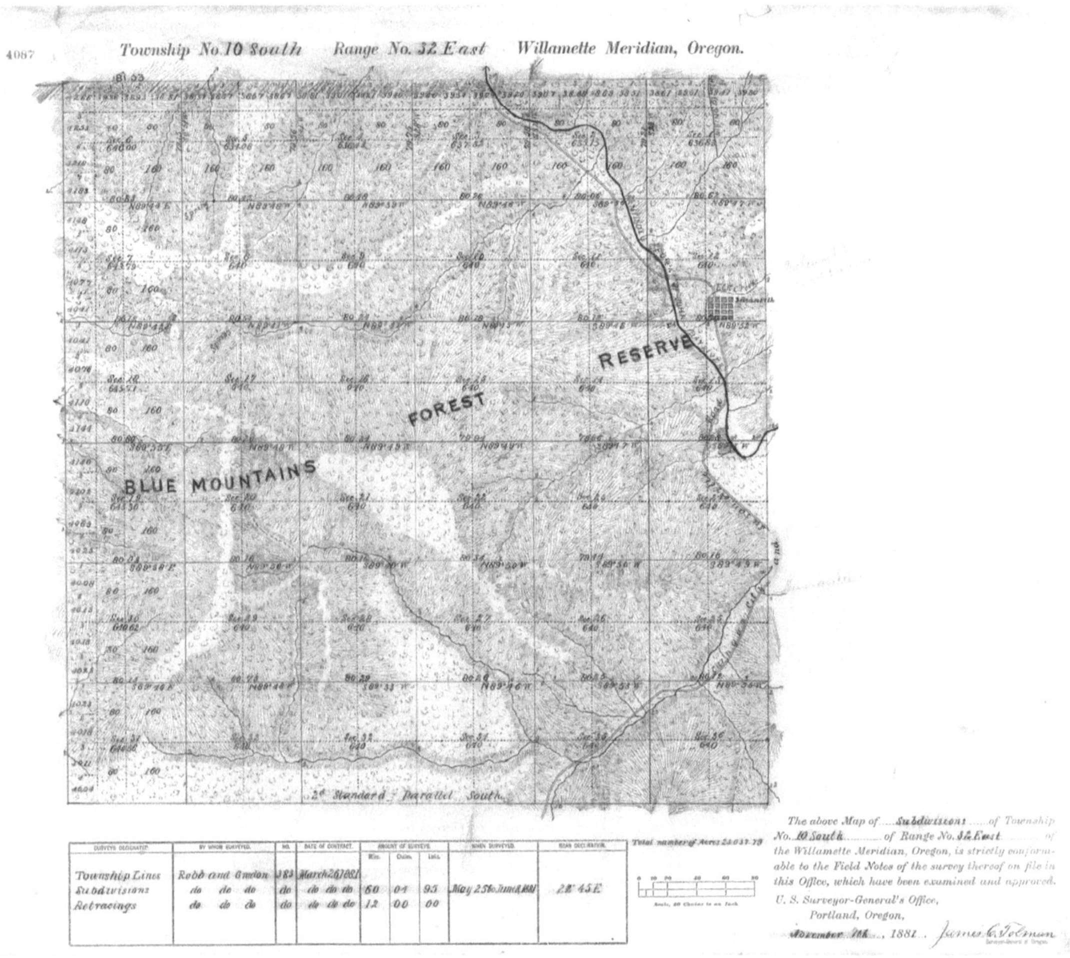

Above and on facing page: General Land Office Cadastral Survey Maps of the Susanville area, 1884. Mining locations were added at later dates. Courtesy of the Bureau of Land Management.

lowest elevation (between 1,000 and 4,000 feet above sea level), along the Middle Fork of the John Day River, the meadowland areas were covered by native bunch grass with cottonwood trees along the riverbanks. At the warm and dry lower elevations (4,000 to 6,000 feet above sea level), ponderosa pine dominated, with comparatively little underbrush. Above the pine areas, the forest took on a more humid aspect and consisted of Douglas fir, western larch, lodgepole pine, and tamarack. At the higher summits (above 6,000 feet), sub-alpine species such as white bark and lodgepole pine, silver fir, and spruce predominated. The understory vegetation throughout the mountains consisted of low shrubs, forbs, and grasses.[1]

The permanent streams of the two districts were fed by snowpack from the upper elevations. The topography of the Blue Mountains shaped the climate of the region. While eastern Oregon generally was semi-arid, resulting

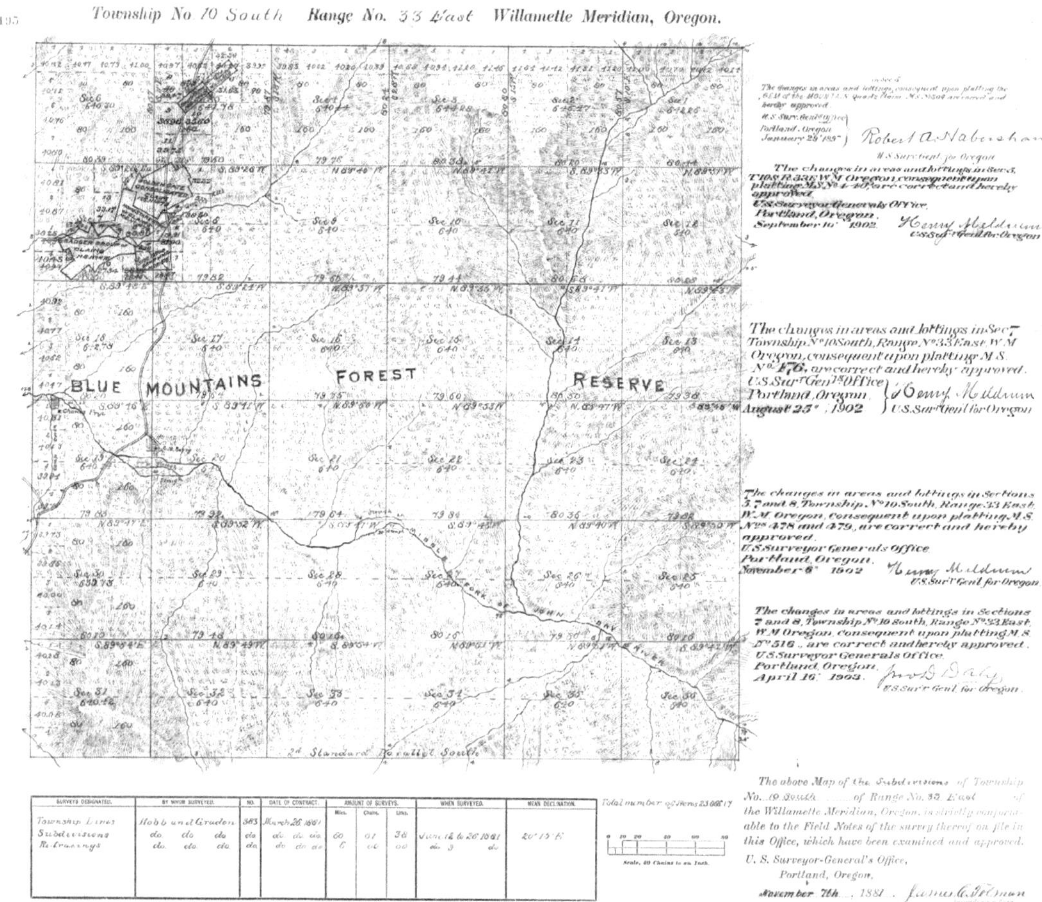

from the rain shadow effect of the Cascade Mountains, the area of the Blue Mountains received more moisture. The mountainous region had an annual rainfall of about 30 inches, although in the highest ranges it could reach as much as 60 inches. Half of the moisture fell as snow. Average temperatures ranged from 20 to 32 degrees in January and 57 to 80 degrees in July. The mountains, canyons, and meadows of the two mining districts supported a variety of animals. The large mammals present included elk, deer, black bears, and cougars. Smaller mammals—such as squirrels, gophers, woodrats, chipmunks, bobcats, beavers, snowshoe hares, and marmots—roamed the region. Many different bird species and fish also resided in the area.

In 1881 and 1882, the General Land Office carried out cadastral surveys of the land in the Granite Mining District (Townships 7S, 8S, and 9S, Range 351/2E, and Range 36E) and the Elk Creek Mining District (T10S, R32E and R33E). This action began the process of turning public land into private property, for settlers or miners could not legally acquire the land until

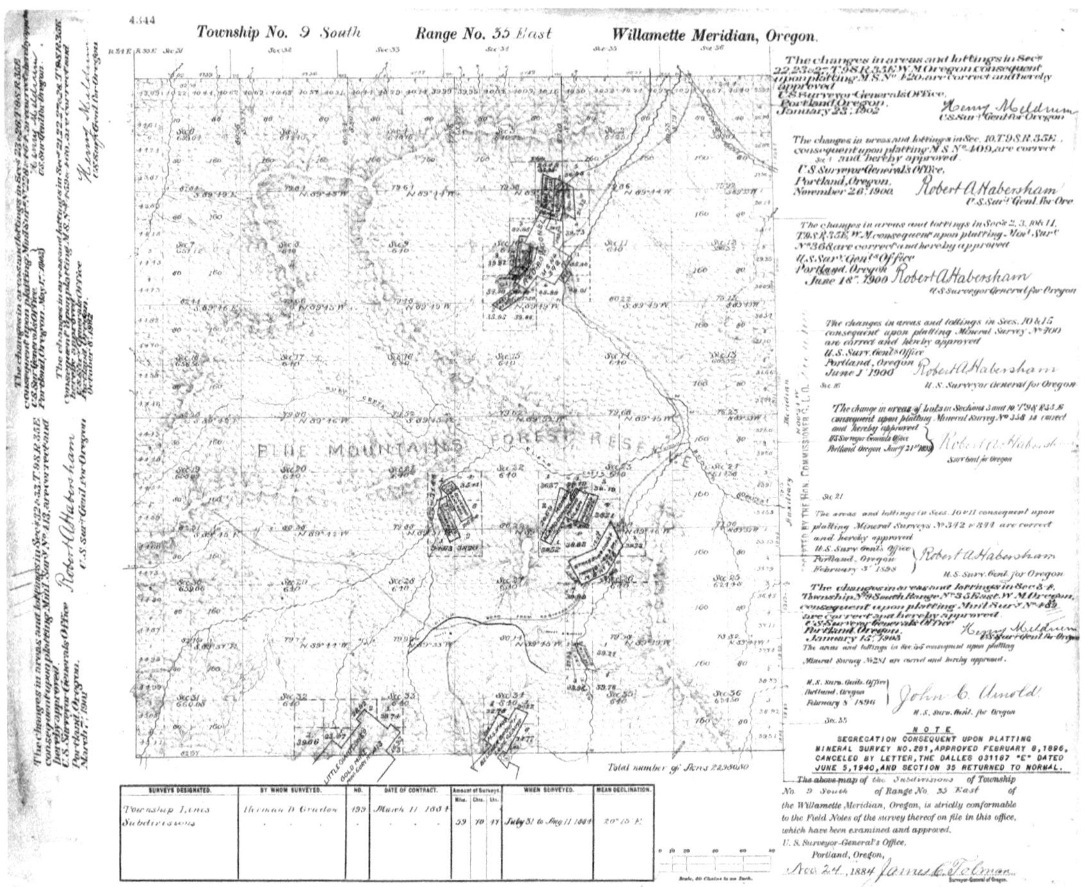

Above and on facing page: General Land Office Cadastral Survey Maps of the Granite area, 1884. Mining locations were added at later dates. Courtesy of the Bureau of Land Management.

the surveys had been completed. By passing the land into the hands of settlers and miners, the federal government intended to help unsettled areas develop and prosper, adding to the nation's wealth. The general description of what the government contract surveyors observed when walking the land showed that the first white emigrants in the region, mostly itinerant prospectors, had made little impact on the landscape. Of the terrain in T8S, R351/2 (containing Granite), for example, the surveyor wrote: "This township is hilly and mountainous, covered with dense growth of timber, mainly Black Pine [lodgepole pine]—is well watered and has some good agricultural land. Is very rich in Gold and Silver, there being many valuable mines now in successful operation. Both quartz and placer."[2] The surveyor of T10S, R32E (containing Susanville) reported: "Land hilly, soil good 2nd rate. Heavy pine timber and dense young pine. Remainder open prairie. Rough, mountainous [terrain]." The surveyor's notes also indicated the location of a store and a dozen dilapidated houses in the southeast corner of Section 12 of T10S, R32E.[3]

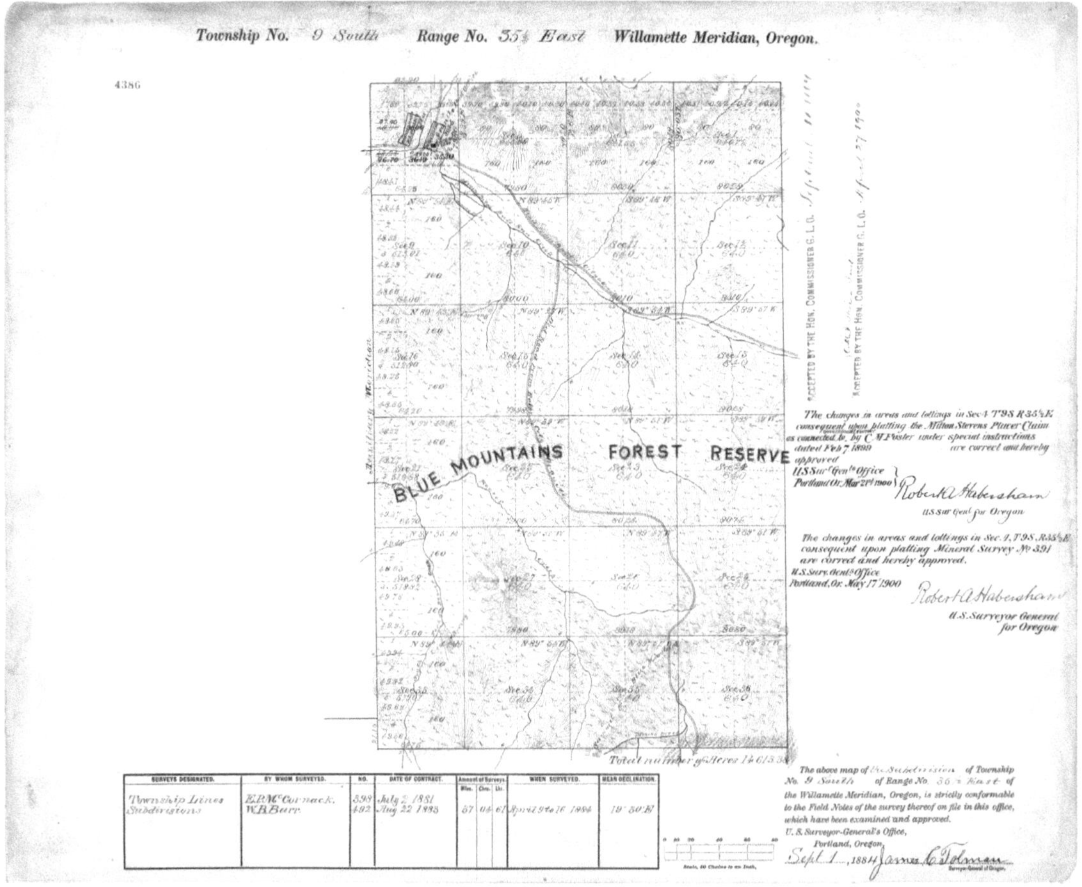

INDIGENOUS PEOPLES

Of course, the white settlers arriving in the 1860s were not the first humans to make use of the resources found in the Blue Mountains. For millennia, Native Americans had lived in or passed through the region between the various forks of the John Day River on their seasonal rounds. The area formed a fluctuating boundary separating the territory of the Northern Paiutes and the homelands of several distinct tribes sharing a common language base known as Sahaptin. The Cayuse, Umatilla, and Tenino represented the latter group—often referred to as Plateau peoples—while the Northern Paiutes considered themselves one people linguistically and culturally. As noted in chapter one, the Blue Mountain region constituted the traditional lands of the Cayuse. The Sahaptin-speaking Plateau groups often established intertribal alliances against their common enemy, the Northern Paiutes from the Great Basin region. When both native peoples attempted to utilize the same portion of eastern Oregon, it could result in fighting when they came into contact.[4]

Over the millennia, the oral traditions of the southern Columbia Plateau people expressed their understanding of the natural world and the fundamental human relationships bound by it. The Plateau groups' subsistence and cultural practices centered on semi-sedentary fishing, hunting, and gathering activities. The Northern Paiute lifestyle, in contrast, consisted chiefly of semi-nomadic hunting and gathering, with plant foods providing much of their diet throughout the year. Both peoples utilized food resources on a seasonal basis whenever and wherever they occurred. Because of environmental differences, the Northern Paiutes emphasized hunting game animals and seed gathering, while the Plateau Native Americans relied more on fishing and root gathering. All subsisted through the winter months on stored dried food.

Some Paiute bands resided permanently in what is now southeastern Oregon, but given the indistinct group territorial boundaries, the various Native peoples shared subsistence areas and engaged in extensive trading. Peaceable relations did not always exist between the Paiutes and Plateau peoples, with frequent raids occurring. Initially, the Plateau peoples had an advantage in these contests, since they had acquired horses before the Paiutes—as early as the mid-1700s. The Paiutes did not possess horses until the mid-1800s. Apparently, neither side sought territory for its own sake, but rather sought advantages in trade and resource acquisition through warfare.

Drastic changes in the Native Americans' lifeways resulted from their first contacts with transient Euro Americans—mostly fur trappers—and, later, from their dealings with white miners and settlers. Initial contacts brought diseases that sharply reduced the Native American population. The acquisition of horses and guns increased Native American mobility and the efficiency of their hunting, while heightening the devastation of their raiding on each other. Over time, white settlers' introduction of domesticated, non-native plants and animals and mining for precious metals altered the natural vegetation and availability of game and encroached on the traditional hunting and gathering preserves of the various Native American groups. The gradual loss of their age-old means of subsistence caused the Native Americans much suffering and ultimately led to steadily deteriorating relations with the influx of Euro Americans.

During the second half of the 1860s, the situation between the Native Americans and Euro Americans became especially troubled. The tribes that used the Blue Mountain region for their seasonal rounds resented the permanent encroachment of the prospectors, miners, expressmen, pack train operators, and settlers. Constant raiding by the various tribes resulted in the theft of horses, livestock, and supplies. In response, the white miners and settlers petitioned state and federal authorities for protection. An early twentieth-century history of Grant County captured the white settlers and miners' negative view of the Native Americans:

> The Piute Indians . . . were warlike by instinct, and marauders from necessity; they knew no peaceful neighbors; every stranger was an enemy. . . . When the whites began to pass through his country as immigrants to western Oregon, or in later years, as gold seekers, the Piutes regarded them as but a new enemy come to torment them when in strong parties, and proper victims to kill and plunder when venturing in small ones.[5]

The United States Army responded to the settlers' complaints by locating a military presence in the Upper John Day Valley at a post called Camp Logan. Established in 1866, Camp Logan garrisoned 40 infantry and cavalry. The Army located the post, which remained active for about two years, six miles south of present-day Prairie City on the north-facing slope of Strawberry Mountain. According to the anonymous author of a history of Grant County, Camp Logan "was most advantageously located. It guarded the best Indian passes into the John Day Valley."[6] It operated in conjunction with other military establishments, such as Camp Harney (1867–1880), and together these posts maintained an unstable peace until the Bannock uprising in 1878.[7]

Established by treaties in 1855 and 1856, the Umatilla and Warm Springs reservations became the homes for various Plateau tribes. After 1868, the Northern Paiutes scattered among several reservations, including the Klamath, Warm Springs, and Malheur in Oregon and others in Nevada and Idaho. Government officials and white settlers selected reservation lands that were presumed unfit for cultivation or mining but compatible with the historic fishing-hunting-gathering lifestyle of the Native Americans. Native and Euro American relations remained fraught throughout the 1860s and 1870s, culminating in the Bannock War of 1878. Though ultimately defeated by the US Army under the command of General O. O. Howard in the summer of 1878, the retreating bands of Paiutes killed two miners near Susanville.[8]

GOLD DISCOVERED AT GRANITE CREEK

The party of prospectors that discovered gold on Granite Creek in the summer of 1862 had their origin in California. Hearing of gold strikes in Boise Basin, Idaho, gold seekers left San Francisco for Portland by steamship and then took an OSN steamboat to The Dalles. At that point, they either procured horses or continued on foot in the direction of Idaho. The argonauts' eastward trek took them along the south side of the John Day River and then up its North Fork to the point where present-day Granite Creek (unnamed at this time) empties into the river. Deciding that this stream offered a better path than the swift and deep river, they followed the creek for four miles and then camped on an open flat with good grass for their pack animals. Traveling four miles farther from this temporary camp brought them to a stream they named Clear Creek. Then, continuing two miles farther up, they reached another stream, which they christened Granite Creek. Here they camped on a heavily wooded field and tested the ground for gold. Finding a good prospect, a portion of the outfit decided to stay, while others pushed on to the Boise Basin. The Granite Creek miners quickly built rude log cabins or set up tents. Most cabins had dirt floors, and cooking had to be done on fireplaces. The nearest supply point was Walla Walla, Washington, nearly 150 miles distant over mountainous and heavily forested terrain.[9]

The miners soon found the location of the initial settlement inconvenient and moved to a new townsite about one-half mile upstream. Since the discovery of gold occurred on July 4, 1862, the miners named the new community Independence. A. G. Tabor, leader of the California prospectors, started the first placer mine at Granite, working 50 or 60 men at a time. When the gold claim gave out, he opened a general store in the community, while his wife operated a boarding house. Tabor, however, never completely gave up mining or other entrepreneurial activities. In 1874, when the townspeople applied for a post office, they had to rename the town, since a post office called Independence already existed in Oregon. They chose Granite for the new name of the settlement, and Tabor became the first postmaster. As placer mining got underway in 1862, the miners adopted local mining regulations for the Granite District, based on those established in Canyon City. The rules specified that each claim should be 200 feet long and the width of the creek. Chinese people were forbidden to own claims in the district, but as early as 1867, Chinese individuals and companies began to purchase or lease mining claims. Some also reworked claims abandoned by white miners.[10]

THE RUSH FOR GOLD

Settlers in the Willamette Valley eagerly sought news concerning the new eastern Oregon gold fields. As early as the fall of 1862, the *Oregonian* reported on activity on Granite Creek, "Several hundred claims have been taken up on the Creek, and all are paying good wages. Companies of three and four men are taking out all the way from $500 to $5,000 per week." The *Oregonian* further noted that a townsite had been laid out on Granite Creek with "some 18 or 20 houses . . . already built, and many more will go up before winter sets in."[11] Even the newspapers in San Francisco took note of the activity on Granite Creek. The *Mining and Scientific Press* reported, on September 11, 1862, that "miners are engaged in fluming, ditching, and other operations, preparatory to taking out large quantities of the covered ore." The following summer (1863), the *Oregonian* stated that some 2,000 miners were reported to be "taking out large quantities of [gold] dust."[12] In September 1865, a correspondent to the *Oregonian* surveyed eastern Oregon mines and noted that the Granite Creek District "has been one of the best paying camps in the country; and promises to return handsomely for an indefinite length of time. The gold of this district is not fine, being largely alloyed with silver. Its market value in Portland is about $14.00 per ounce." The writer went on to note that in this early stage of development, "gold-bearing quartz is found in regularly defined ledges in the neighborhoods of Canyon City, Elk Creek . . . [and] Granite Creek . . . , but there has been no systematic quartz mining done except in one or two places, and only on a small scale in those."[13] According to the *Mining and Scientific Press,* in 1866, Granite Creek continued to yield its gold, as 125 miners worked placer operations that summer.[14] In the mining season of 1868, 250 miners, half of whom were Chinese, labored in the Granite Mining District. The first Chinese miners had arrived by 1867 and began purchasing mining claims that summer, according to Grant County mining records.[15] Placer mining continued at a steady pace in the early 1870s. The *Oregon Business Directory and State Gazetteer, 1873* reported that the Granite Creek placers employed "about two hundred persons, the majority being Chinamen, and mining is carried on throughout the year."[16] A writer to the *Bedrock Democrat* (Baker City) remarked in 1875 that "although [Granite Creek] has been mined for some fourteen years, it still yields its annual tribute of wealth."[17]

In the 1860s, transportation and supply were stumbling blocks in the early development of the mining camp. Not until the 1870s did wagon roads begin to replace the rough trails used by mule pack trains. As the roads improved, freight wagons and Concord stagecoaches were added to the transportation mix. Coaches, depending on size, were pulled by four to six horses, and they carried from six to sixteen passengers. Passenger fares ranged from five to ten cents per mile. Pendleton and Baker City served as the main supply points for the Granite Mining District; each was about 90 miles distant.[18]

By the late 1870s, the best placer mining ground had been worked, and the transition to lode mining had not gotten far. The 1870 federal census of Wealth and Industry provided a snapshot of mining conditions in Oregon at the beginning of the decade. The entire state reported having 168 mining establishments: 165 placer/hydraulic and three quartz mines. These operations employed 880 men, had capital of $321,520, and yielded $417,797 in gold. Oregon's gold output was 3 percent of the total national production, with California representing 58.5 percent and Montana 28.8 percent. The 1870 census also broke down the mining statistics to the county level, revealing that Baker County had thirty-five placer/hydraulic and one quartz mine, employing 135 workers, $207,390 in capital, and producing $107,186 in gold. Grant County, on the other hand, showed better mining results. It had seventy-five placer/hydraulic and two quartz mines that employed 642 men, had capital of $145,400, and produced $246,455 in gold.[19]

During the winter of 1876–1877, a traveler on the route from Pendleton to Granite City described for the readers of the *East Oregonian* the forlorn scene presented by the nearly deserted mining camp along one of the district's many streams: "One will every few hundred yards pass a deserted cabin on Granite Creek . . . where hundreds of men were toiling from morn till night for the precious metal; and where all was excitement and life, is now quiet, deserted; nothing but 'tumble down' cabin remains, the dumb spectator of the past."[20] Over time, news reports from the Granite Creek mining district followed the trajectory common to the other eastern Oregon mining camps during the 1860s and 1870s: Strong returns from placer mining operations for the first half dozen years after the initial discovery and then a gradual decline in output as the surface results gave out. Still, according to US Mint reports, Granite Creek district placers produced $20,000 in both 1882 and 1889. Chinese miners continued their efforts as white miners gave up. Initial attempts at quartz mining, however, did take place in the early 1880s in eastern Oregon. J. W. Virtue, a pioneer quartz mining entrepreneur, reported in 1881 that Baker County had sixty-four stamps in operation at mills in seven quartz mines. At that time, Grant County had forty stamps at mills installed at several different mines.[21]

While the early prospectors also discovered gold-bearing quartz, they lacked the capital, machinery, and technical knowledge to effectively develop this resource. In 1885, a newspaper commentator succinctly described the long-term problem that the eastern Oregon gold fields faced after the early years of mining activity:

> During the ten years following discovery of the mines there was great activity, and nearly three millions of [*sic*] dollars was taken out each year. In 1872 mining operations fell off, and during the 13 years the district has yielded only about one million annually. The causes of the decline . . . were various, but the chief reason was that the shallow placers which could be worked by men without capital, were exhausted. The deep gravel ground which remained, and which still remains . . . can not be worked without capital. There was no capital in the country, the cost of machinery and general supplies was enormous, owing to the isolation of the country, and so the business declined. . . . Most of the mining in this district has been placer or hydraulic but there are quartz mines in great number, comparatively undeveloped as yet.[22]

The struggle to develop the Monumental Mine, a lode discovery outside of Granite, was a case in point.

THE MONUMENTAL GOLD AND SILVER MINE (GRANITE)

The story of the Monumental Gold and Silver Mine in the Granite District reveals the difficult early transition to lode or quartz mining in the gold fields of Eastern Oregon. While several promising ore-bearing ledges were known to exist in the Granite Mining District, the men who located them lacked the capital to work them. Prospectors, as one observer put it, "depend upon making some disposition of them [i.e., claims] to capitalists who can make the necessary improvements in the way of machinery, etc."[23] In the summer of 1875, the *Mining and Scientific Press* reprinted an article from the *Bedrock Democrat* noting that Baker and Grant Counties' miners "want capital that will back the reliable prospector—we want prospectors in whom the capitalists have confidence, and if they are not among us now, there is room here for them, and when they do come and make their reports to capitalists who have faith in their discoveries, then our county will indeed, we think, show its vast richness."[24]

In the spring of 1877, all mining attention focused on the Monumental Mine silver ore-bearing ledge, discovered by Isaac Klopp, a Granite Creek mining pioneer. After discovery in 1873 and initial tunneling in 1875 produced good results, the owners of the claim raised considerable capital to start development but soon realized that their resources were insufficient. These investors sold their interest for $50,000 to new owners from Portland (which included prominent financiers W. S. Ladd, H. W. Corbett, and D. P. Thompson) and Canyon City, who promptly sank $72,000 in a twenty-stamp mill and associated refining equipment—all of which had to be shipped from San Francisco to Portland, brought up the Columbia River to the Umatilla Landing by steamboat, and then hauled ninety miles over a poor country road to Granite Creek. Mining journals could hardly contain themselves in extolling the great opportunity to get rich from the silver mines in Grant County:

> Eastern Oregon, in Grant county, is the next region on the Pacific coast where true fissure veins of silver-bearing quartz of Washoe [Nevada] extent is found, and where silver-producing enterprise will next erect those stupendous and expensive adjuncts of the business. The movement has commenced on Granite Creek, about sixty miles from Canyon City. . . . [W]e are anxious that our own citizens should have a share of those arising values which will pass them to the San Francisco stock exchange if they do not

> take time by the forelock and secure an interest at this time in what is soon to become the northwest bonanzas.[25]

After two years of effort and another $70,000, the mining company completed the mill in September 1879; in addition, it built a three-story hotel, store, sawmill, and other buildings. During the winter of 1879–1880, the Monumental Mine operated full blast, keeping 80 to 100 men—half of whom were Chinese—at work, driving a tunnel 800 feet and processing the tons of ore produced. The mine—at almost 7,000 feet above sea level—with its mill and associated buildings, made quite an impression on early observers. A writer for the *Grant County News* reported on October 11, 1879:

> Arriving at the mill, which is situated on the north side of the mountain ridge and almost 200 yards west of the mine, we were surprised to see a building constructed on the steep mountain side, occupying about 100 feet north and south and nearly 150 feet east and west, and in height 75 feet from the ventilators at the top of the mill to the ground floor of the engine room. . . . [T]here was covering the roof, nearly one hundred and ten thousand shingles. We found the engine of 80 horse power, with two boilers and all set ready for the fire, and as nice and strong machinery as one would wish to see.[26]

Twenty-four years later, the mine still impressed visitors. A reporter for the *Blue Mountain American* wrote, after a brief visit:

> Anyone visiting the mill . . . is compelled at this date [March 7, 1903] to admire the superior work done by the pioneer. The splendid setting of the battery frame, the heavy solid timbers, and the admirable mechanical arrangements of all parts bear the indelible impress of the master builder's hand. Also commanding attention more firmly is the excellent state of preservation of the work. Viewing the mill as it stands now, it requires facts regarding the date of construction to convince one that it has been there for above a score of years.[27]

Another measure of the intense activity at the Monumental Mine can be gleaned from the 1880 census. Apparently, enough of the population lived in the vicinity of the mine that it had a census precinct of its own. In June 1880, the census taker listed a population of 100 residing within the precinct, consisting of ten Chinese and ninety white inhabitants. The white residents included seventy-five men, twelve women, and thirteen children under 18 years old. The women were reported as married and keeping house. The Chinese consisted of seven miners, two cooks, and one launderer. The white workers identified as miners (forty), mining mill hands (six), and non-miners (twenty). The non-miners reported such occupations as carpenter (two), engineer (three), fence builder (six), laborer (one), hotel keeper (two), merchant (one), road supervisor (one), saloon keeper (one), stock raiser (one), teamster (one), and waiter (one). The manufacturing schedules of the 1880 census provided additional information on the operations of the Monumental Mine, confirming the details of newspaper reports. According to the census report, the Monumental Mining Company had a capital investment of $100,000 and employed sixty men at an average daily wage of $4 for skilled workers and $3 for ordinary laborers. The men worked a ten-hour day. The plant had two steam-powered boilers with an eighty-horsepower engine.[28]

The first results from the mine appeared promising as the silver ore was of a good grade, but over time, the veins ultimately proved too small for profitable operation. One government mining report later indicated that the mine only returned $100,000 before shutting down. In May 1882, the *Grant County News* reported that the Monumental Mine had been sold to New York investors and commented, "Let us hope the new company will operate the mine and not upon the share-holders." Over the next twenty years, however, the mine changed ownership several more times and operated only sporadically. A British mining syndicate, beginning in 1888, even tried to make the Monumental Mine pay but failed after heavy expenditures. The British owners invested in the latest equipment, such as Rand drills powered by air compressors. These drills, operated by 30 miners, ran a new tunnel 1,200 feet in length but apparently did not find the hoped-for rich ledge. In the early 1890s, the mine came back under the control of Charles Miller, who had previously operated it. He both ran the mine's mill as a custom operation and continued the search for the mother lode. The low price of silver during the 1880s and 1890s probably contributed to the mine's inability to sustain profitability. It remained mostly idle after 1894.[29]

THE ELK CREEK (SUSANVILLE) DISCOVERY

The discovery and development in the gold fields of the Elk Creek District followed a path like that of the Granite Creek District. In April 1864, prospectors from Canyon City discovered gold on Elk Creek, a tributary of the Middle Fork of the John Day River. The gold-mining camp, located approximately thirty miles north of Canyon City and 125 miles south of Umatilla Landing, soon had a settlement named Susanville. According to a May 9, 1865, report in the *Oregonian*, the camp got its name "in honor of Mrs. Susan Ward, the pioneer female in this new mine." The dispatch went on to describe in excited terms the recent discovery of "the Elk Creek gold quartz lode. . . . One company is now engaged in sinking a shaft, in order to ascertain more satisfactorily as to its richness. . . . Elk Creek is now being worked for a distance of three miles. Placer mines have been found in a flat between Elk and Vincent creeks; but the supply of water is very limited." The initial excitement continued during the spring and summer of 1865, and county officials formally established the Elk Creek precinct with a justice of the peace in April 1865.[30] The miners drafted their mining regulations in July 1865, basing them on the rules established for the Canyon Creek Mining District.[31] In September of that year, the *Oregonian* announced, "These diggings [Elk Creek] have only been discovered a few months but sufficient is already known about them to warrant us in stating that they are quite good. Their extent is not fully known. The camp, however, is quite large. Several thousand persons are already there, and others are going daily."[32]

The mining season of 1866 in the Elk Creek District produced strong results, with a correspondent to the *Mining and Scientific Press* reporting both profitable placer operations and further development of McQuaid's quartz ledge and other mining properties. One miner, the writer noted, even discovered, a few hundred yards above Susanville, a nugget of pure gold worth $203. The correspondent went on to assert "a 20-stamp mill, with the necessary appliances for saving gold, could now get all the work it could do at very reasonable prices."[33] A correspondent to the *Daily Mountaineer* (The Dalles, May 21, 1866) gave a detailed description of the developing Elk Creek mining District:

> The mines on the Middle Fork of John Day are confined to three creeks or gulches, which empty into that stream from the north. Going down stream, they are named, respectively Elk Creek, Deep Creek and Buck Gulch. . . . As a general thing, these diggings are shallow near the mouths of the creeks, and grow deeper as you ascend, until in the upper sections the ground has to be drifted. Some claims, at the mouth of Elk Creek, in the main Middle Fork, have paid well, and are still paying.

This letter also provided the first report of "the advent of a small party of Chinese miners." He added that "these may be looked upon as the forerunners of hords [*sic*] of these industrious people. . . . It is to be hoped that by another year each honest miner in this country will have his dozen *coolies* delving in his claims."

In April of the following year, the news stories continued touting "Elk Creek District . . . [as] one of the richest quartz localities in Oregon." [34] The

Oregonian reported in the summer of 1867, that "parties from Grant County are now in the city for the purpose of purchasing a quartz mill . . . to be located on the middle fork of John Day's river, in the immediate vicinity of rich leads of mineral."[35] The mill, according to later reports, finally went into operation by November 1867; it cost $5,000, but never worked very well and in the early 1870s was moved to another mine in the district. [36] Another glowing account of the riches of the Elk Creek mines appeared in March 1868, again emphasizing the potential for quartz mining, stating the "ledges are mostly lying idle for want of capital to work them successfully."[37] The article went on to note that about 60 persons worked at the Elk Creek mines. Workers also had completed additional ditches that spring to bring more water for the placers.

Rossiter Raymond, the federal government mining expert, visited Susanville and the Elk Creek District in July 1869 and gave a more sober assessment of the mining activity around Elk Creek over the years following the initial discovery. He noted that most of the development work had focused on placer mining:

> The mode of working has been for the most part by drifting under the surface, in the auriferous stratum, and raising the dirt by windlass, water-wheel, or horse-whim. A few claims, however, are worked in the early part of the season, when the snow is melting, by stripping off the surface, fluming, and ground sluicing. The largest nugget found on Elk Creek was worth $480. . . . A ditch commenced in 1865 or 1866 . . . and completed December 1868 . . . brings water to the claims . . . where hydraulic mining is conducted. It is eleven miles long, four miles being two feet wide at the bottom, and three on the top, and seven miles four feet on the bottom and about five on the top, with a depth, on the lower side, of sixteen inches. A large reservoir . . . was completed in 1869. The ditch cost $17,000, gold, and the reservoir about $1,500. . . . A second ditch is in process of construction, about eight and a quarter miles long, two feet wide at the bottom, and three at the top, by fifteen inches deep on the lower side. It will bring water to the Elk Creek diggings, and is now about completed.[38]

Raymond emphasized that while the district contained numerous quartz lodes, little work beyond what was necessary to hold the locations had been done. He also noted that a mining company had erected an eight-stamp quartz mill at the mouth of Elk Creek, and that the machinery proved defective. The mill was idle at the time of his visit. As late as 1871, the mines of Grant County had only this one mill, and even Baker County had only three. According to Raymond, between August 1865 and August 1869, John Blake, a Susanville merchant, had purchased $80,000 of gold dust from the miners of the Elk Creek District. Raymond estimated that the mining season of 1869 would produce about $16,000. He reported that miners' wages amounted to $4 to $6 a day, while board cost $9 to $10 a week. At the time of his visit, the mining district's population consisted of sixty whites and sixty-five Chinese.[39]

In addition to recording the number of inhabitants for Elk Creek in 1870, the census enumerator also reported on industrial activity. This schedule provided additional details on the mining operations at that location: It reported six placer and one quartz operation. Four of these consisted of white concerns, and three were Chinese. The white companies employed thirty-eight miners, while the Chinese had thirty. Two of the Chinese companies employed twelve each, and one employed six. The white mining companies were capitalized for a total of $9,500 and the Chinese at $1,500. The quartz operation had a twenty-stamp mill with an amalgamator. The schedule reported that the white miners had on hand gold dust valued at $16,271 while the Chinese listed their gold dust at $3,625. The census also noted three other industries present at Elk Creek: a steam-powered sawmill plant capitalized at $12,000, a blacksmith shop valued at $1,662, and a bootmaker with equipment worth $250.[40]

In 1873, the *Oregon Business Directory* noted that the Elk Creek district had "yielded well since 1864. Several heavy nuggets have been found there, ranging in value from one hundred to over six hundred dollars." The *Directory* went on to state that "there are two ditches to supply this district with water; one is eleven miles long, and cost nineteen thousand dollars; and the second is eight and one quarter miles in length."[41] Another report by Raymond, in 1874, indicated little change in Elk Creek mining activity: "The middle districts, on the tributaries of the John Day River, with Canon City as their commercial center, may be said to remain about *in status quo*; at least, my correspondents report neither progress nor noteworthy decline." In that same year, an *Oregonian* news item from the Elk Creek mines stated that "there are some 26 well defined quartz ledges, a number of which have been more or less prospected with very good results."[42] Providing more detail on mining in the Elk Creek District in the early 1870s, the *Mining and Scientific Press* reported that John Roy had produced a bar of gold worth $2,000, "the result of a one and a half days' run with an 8-stamp mill crushing nine or ten tons of rock." This was from the Cabell ledge, first discovered in the mid-1860s. The news account concluded, "There is but little doubt but what with proper development, the Elk Creek district will ultimately prove itself to be one of the richest on the coast."[43] At the time, only thirty miners were either prospecting or working on six placer mines in the district. During the next season (1875), Roy continued to work the Cabell ledge with apparent success, the *Oregonian* referring to it as a "truly magnificent mine."[44]

Virtue Mine, eastern Oregon's first, and one of its richest, lode gold mines. Located near Baker City. It operated between 1864 and 1899. Courtesy of the Baker County Library.

THE TRANSITION FROM PLACER MINING TO QUARTZ MINING

During the 1880s, mining in eastern Oregon, and especially in the Granite and Elk Creek districts, followed a similar trajectory. Most work focused on placer operations with limited development of lode prospects. Until the mid-1880s, only three lode mines had operating stamp mills and a serious production of gold or silver: the Virtue (near Baker City), Connor Creek (southeast of Baker City, near the Snake River), and Monumental (near Granite). The Connor Creek Mine, owned by Simeon Reed, a Portland merchant and entrepreneur, was especially rich, yielding at least $1,250,000 between 1880 and 1890.[45]

The Virtue Mine had significance as Oregon's first lode mine and one of the richest producers of gold in the northeastern part of the state. In 1863, W. H. Rockfellow, a freight line operator and part-time prospector, and five companions discovered the lode deposit by following placer gold uphill to its source. The mine was seven miles east of present-day Baker City on barren, low rolling hills. When production began in 1864, the miners worked the course, free gold with two arrastras. Initially, the mine lacked a sufficient water supply. So the first stamp mill was built in the nascent town of Baker City. Later, in 1874, the mine operators found a source of water at the mine site and erected a 20-stamp mill to process the ore from the underground workings. The ore vein of the Virtue Mine extended 1,200 feet in length and reached a depth of 800 feet. From 1863 to 1899, the Virtue Mine yielded $2.2 million (115,211 ounces of gold). Except for a brief time in 1906 and 1907, the mine remained closed after 1899. In the early 1870s, the Virtue Mine was the first lode mine north of California to employ Chinese as underground miners.[46]

As Tables 3.1 and 3.2 indicate, production of gold and silver reached a low point mid-decade; however, by the end of the 1880s, a new dynamic had taken hold. The arrival of the transcontinental railroad in Baker County in late 1884 coincided with a drop off in Nevada and Montana mining, resulting in both a reduction of transportation costs and renewed interest in Oregon gold and silver prospects. A commentary in the April 26, 1884, issue of the *Mining and Scientific Press* concisely summed up the progress of mining in eastern Oregon prior to 1885: "In Oregon and Washington a certain amount of mining development is going on; but as compared to other regions it attracts little attention."

Still, the early 1880s saw sustained mining activity. The remarkable fact was that shallow diggings, ranging from 3 feet to 30 feet in depth, and deeper gravel deposits, from 60 feet to 200 feet deep, worked by placer methods, continued to yield around $1 million in historical dollar values annually in the early 1880s. A federal census bureau report issued in 1882 noted that lode production in 1880 amounted to 23,454.8 ounces worth $190,972, while placer output came in at 62,597 ounces valued at $934,522. In 1881, a knowledgeable observer stated that since 1862, quartz mines had produced about $8 million and that "over three hundred gold and silver lodes are being developed, awaiting capital to put up machinery."[47] In April 1882, the editor of the *Bedrock Democrat* noted, "We are highly pleased to see that many of our placer mine enthusiasts are turning attention to prospecting for quartz and developing the ledges already discovered." Still, he thought, more could be done: "A few years hence many will regret they did not prospect more carefully and extensively their claims."[48] In the early 1880s, Oregon ranked seventh in the nation in gold production and eleventh in silver, and most new settlers in eastern Oregon came to raise wheat, cattle, or sheep. The federal census bureau report of 1882 succinctly summed up the balance between mineral and agricultural production in the state: "Although the mines have become secondary to its agricultural resources in point of importance, they still furnish occupation and profit to many of its inhabitants."[49]

The January 27, 1883, issue of the *Mining and Scientific Press* provided some details on the mining operation in Grant County. After noting that "for

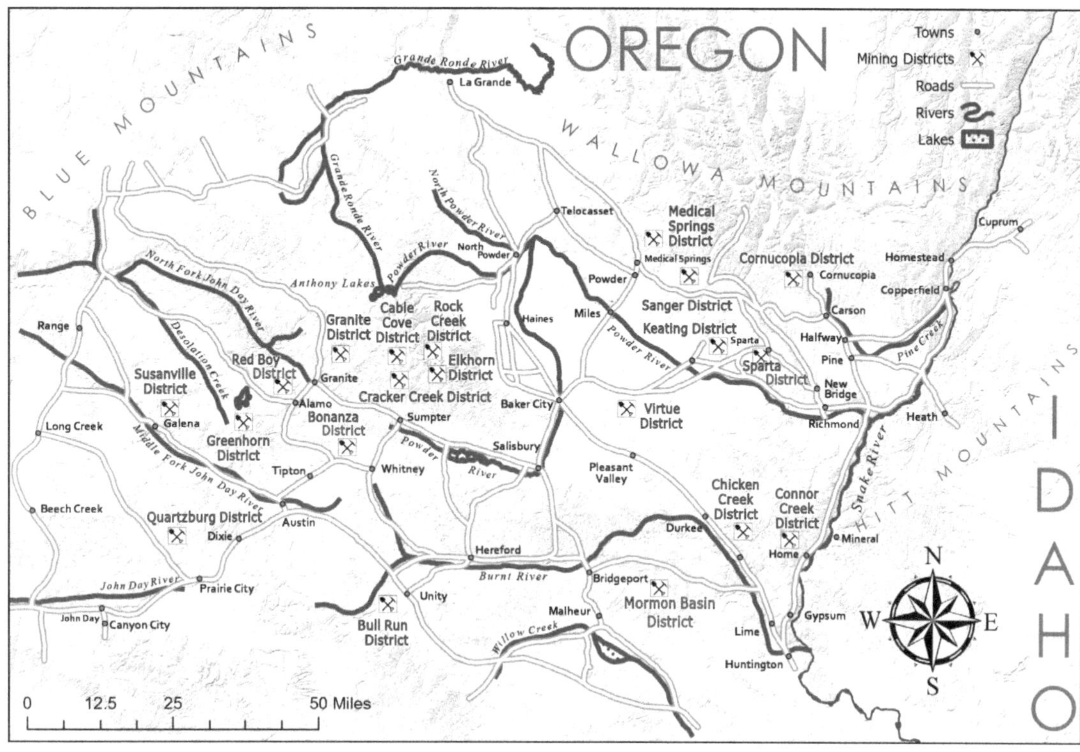

Eastern Oregon mining districts. Map created by Dan Parker (© Oregon State University, published under a Creative Commons BY-NC-SA license)

twenty years a stream of gold has been steadily pouring out of the county," the article listed the total number of men engaged in mining as 497, with 140 of those working in the Granite Creek District and forty at the Elk Creek District. Of those men at either district, Chinese miners predominated: Granite had 110, while Elk Creek had thirty. A measure of mining activity in both districts can be seen in the number of quartz veins and placer claims recorded by the county clerk during 1882. For Grant County as a whole, the clerk listed eighty-three quartz claims and thirteen placer claims. Granite District accounted for twenty-seven quartz and two placer claims, while Elk District had eight quartz and four placer claims. In 1882, Grant County mines yielded $271,000, with Granite District producing $65,000 and Elk Creek District producing $15,000 of the total. In summarizing the work in the Granite District, the article stated, "The famous Monumental Mine, with its splendid mill, is lying idle, but there is said to be reason to believe that it will start up before long. Work is being done upon 20 mines, although little or no rock was crushed during the year. Ore has been worked from 18 mines in the district at various times. . . . Placer mining has been more active on the creek during the year. The Chinese are exclusively engaged in placer mining; the white men largely so. Sixteen hydraulics were operated during the season, and are estimated to have yielded $30,000. All other mines, $35,000." For Elk Creek District, the report noted that "30 . . . Chinese took out $10,000, and a few white men $5,000." The report also listed two promising quartz prospects in the Elk Creek District: The Princess and the Gem of the Mountain. One of the white miners referenced by name was O. P. Cresap, who had engaged in mining at Elk Creek since 1868 and continued to dabble in mining there until his death in 1910.[50]

In its January 19, 1884, appraisal of Oregon's mining activities, the *Mining and Scientific Press* again lamented that "the remote situation of the mines here, the principal districts being far inland and difficult to reach, . . . tended to check mining operations in the state." The article went on to note that the small production of gold in 1883 resulted from a "lack of water for working the hydraulic mines and gulches, from which more than two-thirds of the Oregon gold is obtained." The Chinese now accounted for one-half of the mining workforce but were largely confined to the placer mining of claims abandoned, leased, or sold by white men. In commenting on the nature of Oregon gold, the article stated that most gold-bearing ores contained gold in a free state, but that some were alloyed with silver or contained antimony, which prevented the gold amalgamating with mercury and thus increased the cost of refining. In passing, the article noted that "it has always been more or less difficult to obtain reliable information as to the amount of gold and silver annually produced in Oregon, because so much of the mining is done by Chinese, who are reticent as to their operations, and much of the dust is carried away by private individuals and no record kept." Limited data from the US Mint reports for the 1880s gives the following results from Elk Creek District placers during the 1880s:[51]

1882	$15,000
1883	$25,000
1884	$16,000
1889	$ 8,700

Between 1885 and 1890, the level of eastern Oregon mining activity and outside interest in it increased dramatically. The arrival of the transcontinental railroad decreased the cost of shipping ore for processing and the expense of bringing in heavy equipment needed to extract the below-ground ore. Claim owners now felt incentivized to develop their prospects enough either to sell the claims or entice investors to provide them with the funds necessary to develop the claims into paying mines. The weekly digest of eastern Oregon mining news in the *Mining and Scientific Press* during the last half of the 1880s revealed the steady progression of lode mining improvements at new mines and various mining camps, such as the Cornucopia mines and those in the Pine Creek and Sparta Districts. A *Mining and Scientific Press* writer on May 23, 1885, expressed the newfound optimism regarding eastern Oregon mining: "The signs point to a new era in mining in the county [Baker] and throughout Eastern Oregon. The decline of mining properties in other states and the Territories have sent mining capitalists and prospectors in quest of new fields for investment and work."[52]

By 1887, Eastern, San Francisco, and Portland capitalists began investing in eastern Oregon mines in earnest. For example, Cracker Creek and Granite Creek mines—such as the Eureka and Excelsior, Red Boy, La Belleview, Buffalo, Wide West, and Columbia—moved into active development. The new investors erected stamp mills and extended tunnels to tap the mineral-bearing veins at ever-lower levels. Hundreds of workers found employment in new and old mining operations in the Granite and neighboring mining districts. In particular, the sale in 1888 of the Eureka and Excelsior mines to an eastern investor for a reputed $1 million was hailed as a major advance for eastern Oregon mining. According to the Baker City *Bedrock Democrat*, the purchase would "be the means of attracting other capitalists here."[53] By the summer of 1888, the new owners of the Eureka and Excelsior had shipped in equipment for a forty-stamp mill costing over $100,000 and employed 150 workers. Even the star-crossed Monumental Mine in the Granite District showed renewed activity after its sale to British investors in the fall of 1888. Another measure of the growing investment in lode mining in Baker County was the increase in the number of stamps used to crush the ore holding gold. In 1881, the county had sixty-four stamps; by 1889, it listed 260. Activity had picked up enough in the Elk Creek District that the federal government established a post office in Susanville on September 12, 1888.[54]

By the end of the decade, eastern Oregon gold fields seemed destined for a new burst of development. Lode mining, in particular, appeared to be attracting the attention of men with capital; however, the road to golden riches would not be straightforward. It was all too easy to oversell the prospects for successful mining schemes. A column, for example, in the *East Oregonian* in September 1888, concerning a rush of miners to the Cracker Creek mines near Granite noted:

> The Cracker Creek section is greatly overheated from glowing reports in the Baker City papers . . . [which stated] that the mines were numerous, extensive and undergoing rapid development . . . . On visiting the mines, . . . [it was] found that the bubble had 'busted'. . . . That the mineral belt is reasonably rich is probably true, but reports as to the fabulous wealth have been calculated

> to deceive the multitude, and have succeeded in so doing. . . . Capital and lots of it could probably develop the ledges with profit, but a poor man stands no chance, and this fact is gradually being discovered. The boom is over for the present, at least, as many empty mining cabins attest.[55]

Still, investors with capital could be found if the right incentives existed to entice them.

The creation of the Portland Stock Exchange and Mining Board in 1887 provided an example of the effort of that city's business leaders to make it a growing center of commerce, industry, and capital, partially based on exploiting the potential of mineral wealth. Portland, like many cities of the era, saw a financial exchange as a symbol of a community's growth and prosperity. Between 1885 and 1889, for example, fifteen exchanges started up in eight western states. The operation of the Portland exchange, especially focused on the promotion and sale of mining stocks, sought outside capital, particularly from the East and Europe. The exchange's charter members included twenty-five of Portland's leading businessmen, some with large mining investments in eastern Oregon and Idaho.[56]

The man most responsible for establishing the Portland exchange was Jonathan Bourne Jr. The son of a wealthy Massachusetts family, Bourne arrived in Portland in 1878 at the age of 23 and soon made a name for himself in business and politics. A shrewd, colorful, and ruthless character, he gained admission to the bar in 1881 and quickly became involved in Republican Party politics at both the city and state levels. He served in the state legislature from 1885 to 1886. At the same time, he began investing his inherited wealth in mining and other properties in the Pacific Northwest. He later admitted privately that his obsession with mining caused him to invest "from 1885–93, . . . something over a million dollars in the purchase of mining claims, and the development and equipment of same."[57] As president of the Portland Stock Exchange and Mining Board,

Jonathan Bourne Jr. (1853–1940), US Senator, Oregon politician, and mining investor. Courtesy of the Oregon Historical Society Research Library.

Bourne assiduously promoted and invested in various mining ventures. His efforts to attract outside capital would benefit him personally, while also boosting Portland as a trade center for supplying mines and a possible location for other business and industrial ventures. He was a principal investor in the Eureka and Excelsior mine—a rich Cracker Creek mine.[58]

Table 3.1. Gold and Silver Production in Oregon, 1877–1889*

Year	Gold	Silver	Total
1877	$1,000,000	$100,000	$1,100,000
1878	$1,000,000	$100,000	$1,100,000
1879	$1,150,000	$20,000	$1,170,000
1880	$1,090,000	$15,000	$1,105,000
1881	$1,100,000	$50,000	$1,150,000
1882	$830,000	$35,000	$865,000
1883	$660,000	$3,000	$663,000
1884	$660,000	$20,000	$680,000
1885	$800,000	$10,000	$810,000
1886	$990,000	$5,000	$995,000
1887	$900,000	$10,000	$910,000
1888	$825,000	$15,000	$840,000
1889	$1,200,000	$38,787	$1,238,787

**Historical Dollar Values*

Table 3.2. Gold and Silver Production: Baker and Grant Counties, 1880–1889*

Year	Baker County		Grant County		
	Gold	Silver	Gold	Silver	Total
1880	$286,994	$400	$85,400	$ 543	$373,337
1881	$290,000	$10,000	$280,000	$20,000	$600,000
1882	$250,000	$5,800	$240,000	$25,000	$520,800
1883	$205,000	$2,800	$200,000	$15,000	$422,800
1884	$205,000	$2,800	$200,000	$15,000	$422,800
1885	$355,366	—	$194,600	—	$549,966
1886	$416,765	$9,005	$198,580	—	$624,350
1887	$188,588	$5,153	$163,896	$11,797	$369,404
1888	$205,000	$5,000	$140,000	$10,000	$360,000
1889	$1,038,593	$8,528	$73,989	$9,550	$1,130,660

**Historical Dollar Values. Source: Brooks and Ramp*

STABLE POPULATIONS EXPAND ECONOMIES BEYOND MINING

Even as placer mining began to wane and lode mines seemed unable to reach their full potential without major capital investment, many miners chose to remain in Grant County and take up farming, livestock raising, and/or mercantile business. They settled on the bottom lands of the four forks of the John Day River, in the Long Creek and Fox valleys, and at other locations in northern Grant County. Those who became merchants set up operations in the towns that emerged as supply centers for those who continued to mine and for others who moved into agricultural pursuits. The service centers included towns such as Canyon City, John Day, Prairie City, Long Creek, Granite, and Monument. In 1902, a published history and biographical register of Baker and Grant counties compiled the life stories of prominent male residents, many of whom had arrived prior to 1870. Of the 262 men profiled, eighty-four (32 percent) initially came to Baker County as miners, while in Grant County, thirty-five of ninety-four (37 percent) men sketched in the history volume started as miners. Even though most eventually took up other occupations, many still engaged on the side in mining activities, particularly as investors.[59]

A few examples from the Elk Creek and Granite Mining Districts illustrate how mining fueled the settlement and growth of eastern Oregon. George Rader arrived in Canyon City in 1862 and spent the next ten years working mining claims there and at Susanville. In the mid-1870s, he began running a few head of cattle in Long Creek Valley and soon took up a homestead patent in the valley. Over the next twenty-five years, he added thousands of acres to his holdings and raised both cattle and sheep. Rader became one of the largest taxpayers in Grant County before his death in 1919. Another pioneer miner at Elk Creek, John Blake, began mining in 1864 and then used his mining returns to open the first general merchandise business in the Susanville mining camp. He successfully ran the store until 1889, when he disposed of it and bought into a mercantile firm at John Day. Blake still held mining property (the Gem Mine) in the Susanville District until his death in 1906. As noted above, Oliver P. Cresap, another Canyon City pioneering miner of 1862, moved to Susanville in 1868 and mined there for thirteen years. Then, between 1882 and 1895, he served two terms as county sheriff and ran a general store in John Day. Cresap continued to prospect and mine at Elk Creek until his death in 1910. The Granite Mining District offered the example of A. G. Tabor, who was with the initial party of prospectors at Granite Creek in 1862. While keeping a hand in placer and lode mining until his death in 1894, Tabor also ran a general store and served as the first postmaster in the town of Granite. His son, James, continued to reside there off and on, engaging in mining and mercantile businesses until his death in 1948.[60]

A correspondent to the *East Oregonian* (Pendleton), writing from the town of Long Creek in northern Grant County, humorously described the relationship between ranching and mining in the 1880s:

> Many people in Oregon . . . are not posted in regard to mines and mining on the North and Middle Forks of the John Day's rivers. It is a fact that placer mining is carried on here year after year, and is a source of revenue to Oregon, and especially to the settlements around the mining district. One will hear on the streets: 'Hello old son, where did you come from?' 'Just come from my ranch.' 'Where are you going with those pack animals?' 'Going up the Middle Fork [of the John Day River] to my claim.' 'How did you make it up there last summer?' 'Well, she panned out tolerably well considering that the water gave out unusually early. Jim and I cleaned up $800. . . . I lost part of my cattle the last of February, and Jim lost about half of his sheep; so you see we have to get down to bedrock, and save all the stuff this spring. Come, go along, there is a main lead somewhere above there and will be lots of prospecting done this summer. You may find a pocket, and can get a color most any place. . . . Come, old pard, and hit the trail a lick for luck.' You can hear such conversation here every day, and see men going in to work their claims, or to prospect.[61]

The steady economic and population growth of Grant and Baker counties during the 1880s was reflected in the federal census data. Between 1880 and 1890, Baker County grew from 4,616 to 6,764 (a 47 percent increase), and Grant advanced from 4,303 to 5,080 (an 18 percent increase). This progress occurred despite both counties losing almost two-thirds of their southern land mass to the formation of two new counties. Baker County, moreover, lost 2,601 of its population to Malheur County and $859,624 of its taxable property; and Grant County, 2,559 to Harney County and $1,618,220 of its taxable property. The state legislature had carved Malheur out of Baker County in 1887 and Harney from Grant County in 1889. Table 3.3 details aspects of the agricultural progress of Baker and Grant counties.[62]

By examining the precinct level of the federal census schedules, it is possible to create a profile of the basic demographic features of the Granite and Elk Creek mining districts in 1870 and 1880. At a minimum, the data revealed the sex, ethnicity, age, mobility, and occupations of the inhabitants of those communities. In addition—depending on the census—the schedules listed other items, such as literacy, citizenship, employment, and certain economic data. Together, such information made it possible to create a profile of the

Table 3.3. Agricultural Production, 1880–1890

	1880		1890	
	Baker County	Grant County	Baker County	Grant County
Farms	453	306	455	624
Improved Acres	49,949	30,407	60,753	84,215
Value of Farms	$538,350	$458,675	$2,080,700	$1,683,000
Wheat (bu.)	33,956	45,892	56,739	26,479
Oats (bu.)	103,816	35,206	61,154	19,312
Sheep	30,652	58,490	19,435	237,346
Horses	9,975	10,286	9,723	13,447
Cattle	77,893	78,364	18,837	21,967

Source: US Census Schedules

mining settlements and, by comparing such information from other mining communities throughout the West, determine what is typical or atypical about Granite and Susanville. Table 3.4 displays the population of the two precincts in 1870 and 1880. Elk Creek and Granite census precincts accounted for 45 percent (425 out of 940) of the Chinese in Grant County in 1870 and 28 percent (251 out of 905) in 1880.

Table 3.4. Population of Select Mining Communities, 1870–1880

	1870			1880		
Precinct	Total	White	Chinese	Total	White	Chinese
Granite	448	83	365	200	77	123
Elk Creek	103	43	60	150	22	128

Source: US Census Schedule

When the census taker arrived at the Granite precinct in July 1870, he found a mining settlement strung along the banks of Granite Creek that exhibited elements of permanence. Placer miners in the vicinity of Independence had been working the several streams that feed into the North Fork of the John Day River for eight years, and a nucleus of support businesses had set up shop on the high ground above and slightly east of Granite Creek. The miners initially called this settlement Independence, but it would soon be renamed Granite once a post office opened. The occupations held by white males as listed in the census indicated that while forty-eight were miners (76 percent), fifteen (24 percent) had other jobs: blacksmith (three), butcher (two), dairyman (one), huckster [peddler or small trader] (three), merchant (two), sawmill worker (two), and teamster (two). The birthplaces of the white inhabitants revealed that sixty-four were born in the United States and only eighteen in Europe or Canada. The precinct had twenty-four white households, averaging

3.5 persons each. Four of the white households consisted of nuclear families, averaging three children each.[63]

From the census data, the key demographic characteristics of the precinct in 1870 denoted a community of young, single males, with a large Chinese presence. The census enumerator recorded a total population of 448, containing only seventy-one white males (15.8 percent) and twelve white females (2.7 percent). The enumerator listed six of the seven women over 18 years of age as "keeping house" and the other one as a "hotel keeper." (Her husband was a miner, and she had three children at home.) The average age for white males was 32.45 and for white females, 19.25. Strikingly, Chinese males accounted for 365 (81.5 percent) of the total population, while white inhabitants (male and female) numbered 83 (18.5 percent). The youngest Chinese male was 12 years of age, and the oldest, 65, with an average of 31.12 years. There were no female Chinese present. While 337 (92 percent) of the Chinese worked as miners, twenty-six (8 percent) held other occupations: butcher (one), cook (three), doctor (one), gaming house operator (eleven), merchant (nine), and tailor (one). The precinct had forty-three Chinese households, averaging 8.4 persons each. The Chinese set up their camps to the west and below the village of Independence, along the banks of Granite Creek.

When the census taker appeared in the Granite precinct in June 1880, he encountered a community with a slightly different racial balance than that of 1870. The Chinese now accounted for 61 percent of the population, and whites, 39 percent. The total population consisted of 171 men (85 percent), thirteen women (7 percent), and sixteen children (8 percent). Chinese males (with an average age of 34.33) and one Chinese female (age 25) numbered 123 of the total population, while fifty-six white males (with an average age of 34.16) and twenty-one white females (with an average age of 18.14) amounted to 77. The census listed the lone Chinese female as a house servant. The Granite precinct had twenty-nine white households, averaging 2.7 persons each. Thirteen of the white households consisted of nuclear families, averaging 1.2 children each. The eighteen Chinese households averaged 6.8 persons each. Of the white inhabitants, seventy-one reported being born in the United States, and only six in Europe or Canada.

Just as in 1870, the 1880 census showed that mining represented the chief occupation in the Granite precinct. Fully 98 percent (119) of the Chinese and 83 percent (thirty-six) of the white males worked as miners. The non-miner Chinese consisted of two cooks and one storekeeper. The thirteen non-miner whites (17 percent) held a variety of jobs, including three butchers and one each of a blacksmith, boarding house proprietor, carpenter, fence builder, laborer, livery stable proprietor, merchant, saloon proprietor, stock raiser, and wood chopper. The Granite community demonstrated little persistence over time, as only six white inhabitants from the 1870 census also appeared on the 1880 list.

When the census taker reached the mining camp of Elk Creek in August 1870, he encountered a racially mixed population of young, single males. The makeup of the precinct's inhabitants reflected a placer mining camp receding past its prime. Six years of intensive mining had skimmed the easily recovered gold from the creeks and gulches of the Elk Creek district, and white miners had begun to give way to Chinese gold seekers. The census enumerator recorded a population of 103, consisting of forty-three white and sixty Chinese. Thirty-four males and nine females made up the white population while the Chinese contingent counted fifty-eight males and two females. The Euro Americans lived in fifteen separate households, averaging 2.9 persons each. The Chinese lived in eight individual households, averaging 7.5 persons each. Two of the Chinese households had twelve each. The census listed four white nuclear families, three of those having children present.

The census enumerator identified two of the Chinese households as "gambling houses," and each had a Chinese female—ages 20 and 24, respectively—as a member. The older female was listed as the head of her household. Except for nine males, all the white inhabitants were born in the United States. Only one male adult had been born in Oregon; the others hailed from a variety of states. All the Chinese had been born in China. The average age of the white and Chinese males in the camp was almost the same: 32.82 for the former and 33.6 years for the latter. The average age of the white females was 21.88. The census roll listed fifty-six of the Chinese as miners and two as gamblers. The white males had more diverse occupations: nineteen miners, four mule packers, and one each as butcher, dairyman, trapper, blacksmith, merchant, farmer, and teamster.

The 1880 census for the Elk Creek precinct revealed a community with an even more skewed racial profile than that in 1870. In 1870, Chinese people accounted for 58 percent of the population (sixty), but ten years later, that group had grown to 85 percent (128). Correspondingly, the white numbers had declined from 42 percent (forty-three) to 15 percent (twenty-two) of the total. Overall, the total precinct had expanded by forty-seven individuals over the decade (103 to 150), an indication of renewed mining activity. The census enumerator counted seventeen white males, five white females, and 128 Chinese males. The Euro Americans lived in eight separate households, consisting of 2.8 persons each. The Chinese lived in sixteen separate households, averaging eight persons each. The census listed three nuclear white families, two of those having a child present. There were no female Chinese in the precinct. None of the white male inhabitants were born in Oregon. Two came from Prussia, and the rest hailed from other states. All the Chinese had been born in China. The average age of the white and Chinese males in the camp diverged slightly more than it had in 1870: 37.18 for the former and 33.63 for the latter. The five Euro American females had an average age of 24.4. As a group, the white miners were slightly older in 1880 than was the case in 1870, but the Chinese had the same average age in both censuses. The census reported that 122 of the Chinese worked as miners while the remaining consisted of five cooks and one carpenter. The white males reporting occupations comprised a small group: one merchant, eight miners, three farmers, and one laborer. Only two white males listed in the 1880 census had been present in 1870: a merchant and a miner.

To date, historians have provided few systematic demographic studies of nineteenth-century western mining communities. What has been done has focused chiefly on Nevada (Comstock) and California (Grass Valley). Those studies showed a high percentage of foreign-born people in 1870: 44.2 percent in the Comstock district and 75 percent in Grass Valley. The existing work on western mining camps also emphasized, based on anecdotal evidence, that such places in California, Nevada, and Colorado largely consisted of young, single, transient males from diverse social and economic backgrounds. By the time the federal census takers reached Granite and Elk Creek (Susanville) in 1870 and 1880, the initial rush had long since passed, and the inhabitants of those mining communities reflected a different make-up than the one suggested by the traditional view. In 1870, the white population of Granite and Elk Creek was composed largely of single, native-born, and relatively young (an average age of 32) males; in 1880, the average age for males in the two communities stood at 35. Only 22 percent (eighteen) of Granite's white inhabitants in 1870 had been born outside of the United States. Elk Creek's percentage of foreign-born was almost identical: 21 percent (nine). By 1880, the white, native-born population had increased to 92 percent (seventy-one) in Granite and 91 percent (twenty-two) in Elk Creek. Typical of other western mining locations, the 1870 and 1880 census records for Granite and Elk Creek indicated little permanence in the population over time.[64]

Canyon City, Oregon, ca. 1880, as seen from the north, looking south toward Little Canyon Mountain. Heavy-mined Canyon Creek fans out on the valley floor. Courtesy of the Grant County Museum.

Chapter Four
Lode Mining Takes Off

The 1890s began a new era in eastern Oregon precious metal mining. While placer mining continued at a steady pace, mine owners now ramped up lode operations. The arrival of the transcontinental railroad in the late 1880s provided the previously unavailable, cost-effective transportation to bring in the heavy equipment needed for underground work and the shipping out of ore for processing. The railroad also enabled investors or their representatives to inspect mining properties more conveniently for potential purchase and development. In addition, recent scientific experiments had led to improved processes for extracting the gold from its ore matrix, making mining more profitable. All these advances in mining operations encouraged renewed interest and investment in eastern Oregon mines. The new lode mining, moreover, was typical of gold fields around the world in the late nineteenth and early twentieth centuries. Gold rushes in Australia, Canada, South Africa, Ghana, and Fiji all relied on corporate investment, sophisticated milling and refining technology, and wage labor to operate successfully.[1]

LODE MINING TAKES OFF

A sense that eastern Oregon gold fields were attaining new life in the late 1880s could be seen in such comments as appeared in the August 1, 1889, issue of *The West Shore*. In writing about the recent growth of Baker City, the magazine noted:

> Within the past year or two more work that shows results has been done in the Baker City country than during any previous

> decade. In fact, it is but a short time since the minerals of that section began to attract the attention of capitalists from abroad. The capacity of old quartz mills is being enlarged and new mills are being erected in every direction. Capital is flowing into the country faster than ever before, and there is very pronounced activity in all kinds of mining property, including preparations for getting out precious metals from some of the richest deposits in the west.

Six months later (January 11, 1890), the editors of *The West Shore* once again noted the renewed mining excitement in the Baker City country. They reprinted a speech given at the Board of Trade of Baker City, which detailed recent mining developments and summed up with the following assertion:

> Take a retrospective glance over the past five years, and see what has been accomplished in our mineral development. When the railroad was completed, our mines had received no attention from capital. We were a terra incognita to the adventurous speculator, and they both turned from us with contempt or pity when we mentioned our mines. All this is now changed. Capital now seeks investment in our mining properties and the ever necessary middle man and promoter is always with us. We have great mineral belts traversing our country showing more wealth, and promise of more wealth, than the famous Comstock has produced. Mining men from all section of America and England have visited us, and their universal opinion is that our county and country has no superior on the globe.

The speech concluded with a rhetorical flourish about Baker City's golden future: "Then what may we not hope to see in the next five years, now that capital and enterprise have both joined hands to make our capabilities known?" The speaker answered his own query with the clarion call: "We have the natural resources. Let us have the faith and energy to do it."

Gold and silver production figures for the late 1880s and the first half of the 1890s dramatically bear out the renewed state of eastern Oregon mining. For example, between 1885 and 1889, Baker and Grant counties produced an annual average of $606,876 of precious metal, while between 1890 and 1894, their annual output more than doubled, averaging $1,398,137. By the early 1890s, the eastern Oregon production of gold and silver came from eight mining districts surrounding Baker City in Baker County and five districts northeast of John Day/Canyon City in Grant County. In January 1890, one mining expert counted at least nineteen mines, which operated ore processing mills containing 255 stamps, in the Baker City gold belt.[2]

In its July 5, 1890, issue, the Baker City *Bedrock Democrat* listed the new investments and developments in the mines located in the Cracker Creek, Granite, Cable Cove, and Cornucopia districts. The mines receiving the greatest attention included Eureka and Excelsior, La Belleview, Baisley-Elkhorn, Robbins-Elkhorn, and Virtue. The newspaper confidently opined, "It is plainly apparent that our mines are forging ahead prominently." Even placer mining saw renewed activity in the Granite District, promising "good returns," according to the *Bedrock Democrat* of September 26, 1891. In October 1891 alone, Baker City's two banks and Pacific Express Company shipped at least $100,000 (the historical dollar value) in gold dust and bullion.[3]

In the early 1890s, the new Greenhorn Mining District, immediately to the west of the Granite District, saw initial development, even though valuable deposits of gold had been discovered there as early as 1870. At least sixteen claims were being actively worked, and the mine owners pressed Grant County to improve the wagon road to the Greenhorn camp. One mining enthusiast wrote to the *Oregonian* on September 12, 1891, that "the Greenhorn mining district in Grant county . . . [was] believed by many to be one of the richest mining regions in the United States." Quartz deposits from the various lodes claimed to assay between $18 and $6,000 per ton. The isolated Greenhorn mines were about sixty miles south of Pendleton, Oregon, and forty-four miles west of Baker City, the nearest railheads.[4]

All the new mining activity required heavy equipment, and the April 23, 1892, issue of the *Bedrock Democrat* excitedly noted the arrival via railroad of carloads of mill machinery from Chicago and San Francisco foundries, then taken by wagon to various mines. For instance, the North Pole Mine in the Cracker Creek District alone received 200 tons of machinery in October 1892. In addition, mine owners tried some of the new methods for treating ores, such as the chlorination process, at the Eureka and Excelsior Mine in the Cracker Creek District and other mines in the Cornucopia District. The *Bedrock Democrat* thought the success of the new treatment methods meant "the dawn of a new era for base ore mining in Eastern Oregon."[5] Even as mining experts touted the richness of eastern Oregon mines, they continued to plead for ever more capital investment. As one correspondent from eastern Oregon wrote to the *Mining and Scientific Press*: "True there has been lots of money spent in eastern Oregon in the way of [ore process] mills and 'wildcat' mining schemes, but bring up all monies spent in mining projects, and Eastern Oregon will pay as many mills on the dollar, for capital invested, as any other state. And we can furnish as many big mines in eastern Oregon as any State, according to development work."[6]

Enthusiasm for eastern Oregon mining continued in 1893. A report on gold mining appeared in the *Mining & Scientific Press* in September of that year, which asserted that "the mines of Oregon share in the recent general movement of gold properties," and that "around Baker City, in eastern Oregon, the prosperity of the county almost assumes the dimensions of a 'boom'." The following year, moreover, witnessed the second-highest yield of the decade for the mines of Baker and Grant counties. During 1894, the gold and silver product amounted to $1,647,270 in historical dollar values; only 1891 exceeded that sum, coming in at $1,849,131 (see Table 4.1). Again, both placer and lode mines accounted for this result.

Table 4.1. Gold and Silver Production: Baker and Grant Counties, 1890–1899*

Year	Baker County		Grant County		
	Gold	Silver	Gold	Silver	Total
1890	$735,000	$127,540	$90,000	$129	$952,669
1891	$1,499,014	$221,333	$124,487	$4,297	$1,849,131
1892	$1,121,302	$5,157	$53,780	$40	$1,180,279
1893	$1,149,184	$13,500	$198,650	—	$1,361,334
1894	$1,507,066	$10,351	$129,853	—	$1,647,270
1895	$1,087,283	$10,963	$101,853	—	$1,200,099
1896	$1,100,000	$20,000	$100,000	—	$1,220,000
1897	$1,008,440	$86,159	$86,969	$4,880	$1,186,408
1898	$818,269	$110,506	$143,463	$32,769	$1,105,007
1899	$696,560	$74,884	$217,054	$86,626	$1,075,124

**Historical dollar values. Source: Brooks and Ramp*

By October 1894, the *Bedrock Democrat* was boasting that "more development work has been done on the mining properties during the present year of 1894 than at any other time in the history of Baker County, and the work thus prosecuted has been attended with good results in placing before the eyes of the outside world properties that will pay big dividends on the capital invested in their operation. Men of wealth are becoming interested and taking hold of our mines with the intention of erecting works for the proper treatment of the ores." The newspaper article concluded by predicting that "the great amount of development work done on mining properties the past spring and summer will . . . be amply rewarded during the coming spring and summer of 1895 by the investment of a large amount of capital by men of wealth . . . who are eager to invest in developed properties."[7]

The pace of mining activity accelerated in the second half of the decade, reinforcing the trends of the first part of the 1890s. Out-of-state investors increasingly bought, bonded, or leased lode and placer mines and invested

An 1890 Union Pacific Railroad map of the state of placer and lode gold mining on the eve of a major expansion in eastern Oregon. Compare with map 5.9. Courtesy of the OHS Research Library.

large amounts of capital in equipment and infrastructure. For example, during 1897, the Bonanza Mine (Greenhorn District) changed hands for $750,000, the La Belleview Mine (Granite District) sold for $125,000, the Ramshorn Group of Mines (Baisley-Elkhorn District) went for $75,000, and the Baisley-Elkhorn (Baisley-Elkhorn District) marketed for $65,000. In all, that year, sixty-four mining properties underwent an ownership change for a total expenditure of $1,237,000. For all the financial activity, however, production did not show an increase over the course of the decade. According to US Mint records, 1894 marked the peak of production in eastern Oregon ($1.6 million), while 1899 represented the low point ($1 million). The average output for Baker and Grant counties between 1895 and 1899 was $1.2 million. Gold production for the entire state of Oregon tracked the eastern Oregon numbers. In 1894, the state total reached $2.1 million, while in 1896, it came in at $1.3 million, and the average for the last half of the decade stood at $1.5 million (see Tables 4.2).[8]

The perennial lament of miners, as reported in the press, revolved around the need for ever more capital investment. A correspondent to the *Oregonian* at the beginning of 1897 lamented, "The drawback to the industry of gold mining in the districts adjacent to Baker City has always been and is today want of capital." The high cost of development limited local efforts to get

Table 4.2. Oregon Gold and Silver Production, 1890–1899*

Year	Gold	Silver	Total
1890	$1,087,000	$129,199	$1,216,199
1891	$1,994,622	$296,280	$2,290,902
1892	$1,491,781	$64,080	$1,555,861
1893	$1,690,951	$13,557	$1,704,508
1994	$2,113,356	$10,315	$2,123,671
1895	$1,837,682	$15,192	$1,852,874
1896	$1,290,964	$71,811	$1,362,775
1897	$1,354,593	$109,643	$1,464,236
1898	$1,216,669	$165,916	$1,382,585
1899	$1,467,379	$187,932	$1,655,311

**Historical dollar values. Source: Brooks and Ramp*

mines in shape for sale. As the correspondent phrased it, "Men have come into this section who, to some extent, are practical miners, but they have failed to handle the ores as they should be handled to produce satisfactory results." He concluded by arguing that if eastern Oregon mining succeeded, then Portland would benefit as well: "What is needed is capital and a smelter and when intelligent methods are adopted, Portland will find her resources taxed to supply the demand for the articles in which her business men deal."[9]

By the middle of 1897, a comprehensive survey of the eastern Oregon gold fields written for the *Mining & Scientific Press*, optimistically concluded that "in every instance the mines show a good quality of ore at the lower levels, and all those who visit the region will be convinced that the mines are permanent and may be expected to continue to the limits of mining economy." [10] The author based his conclusions especially on the mines adjacent to Baker City and those in the Cracker Creek and Granite districts, such as the Virtue, Eureka and Excelsior, North Pole, Columbia, Bonanza, and Red Boy. These mines were reputed to produce between $20,000 and $30,000 per month of product. In the following years, other reports on the Cracker Creek mines noted that increased use of chemical concentration to separate the gold from its complex, sulfide matrix was underway, especially a chlorine-bromine process at the Golconda Mine, while the North Pole Mine relied on a cyanide concentration plant. The new Golconda plant was reputed to have cost $100,000.[11]

Even the railroads got into the act of mining promotion. To generate traffic over their line, the Oregon Railroad and Navigation Company periodically published promotional tracts aimed at prospective settlers from the eastern part of the country. In 1897, it issued a pamphlet and map entitled *Gold Fields of Baker County, Eastern Oregon*, averring that the settlers coming to Baker

County would find "its every peak and foothill . . . ribbed with royal ore, and its every gulch and creek-bed a deity-made deposit of golden sands . . . . It has produced untold millions of dollars in gold."[12] The author assured his readers that the mines around Baker City offered plenty of work, employing between 1,000 and 2,000 men, earning at least $65,000 per month or $750,000 a year.

Over the course of the following year, eastern Oregon mining investment apparently did pick up. A field assayer for the US Mint reported in June 1898: "In the northeastern part of the state [Oregon] the exploitation of many meritorious claims was vigorously prosecuted last year and a number of new properties are included in the list of producers. Unusual attention is turned to this section, and activity characterizes the entire industry. New machinery is contracted for to work ore heretofore only under development, and a large and increased output in the future is assured."[13] Significantly, foreign investment also poured into eastern Oregon in the 1890s. One cursory study by a historian found that between 1888 and 1900, British investors incorporated nine joint stock companies with a nominal capitalization of $3.3 million to operate mines in Grant and Baker counties.[14] This foreign capital investment represented but one example of the process that historian William Robbins has called the capitalist transformation and industrialization of the American West as natural resources were opened to development. Eastern Oregon mining was becoming part of the globalization of markets as capitalist investment expanded worldwide.[15]

In particular, mines in the Bourne District (Eureka and Excelsior, North Pole, and Columbia), Granite District (Buffalo and La Belleview), and Susanville District (Sloan and Haskell Placers and Badger Mine) received notice, while a new district also gained increasing attention: Greenhorn (Red Boy and Bonanza). Another observer, writing in 1899 for the *Mining and Scientific Press,* supported the US Mint's field assayer's opinion: "The good record and success of a number of mines [near Sumpter] . . . . have established confidence and imbued the district with enthusiasm, the result of which is seen in the purchase of a number of prospects by Seattle, Spokane, Chicago and Montreal men; and the activity displayed by the newcomers in pushing development work and building mills is evidence of substantial backing."[16]

While recognized as early 1891, a Granite, Oregon, correspondent to the *Mining & Scientific Press* wrote in January 1895, concerning the Greenhorn area: "The Greenhorn mountains are undeniably rich in minerals, and it is only a question of time when development will demonstrate the existence of ore bodies of incalculable importance to the mining world." The writer went on to boast that "nature could not have done more in the shape of natural advantages for the successful working of the properties, even though Providence has taken especial efforts to do so." After enumerating some of the prospects in early development, he closed his Greenhorn panegyric with a grand flourish: "The entire country is picturesque; the mountain scenery is embellished here and there by beautiful lakes filled with the finest trout, while every direction little streams wind their way, all of which are well stocked with speckled beauties. Game of every description natural to this latitude is plentiful. The climate is delightful, and viewed impartially, it is undoubtedly one of the most interesting mining camps on the continent."[17]

A report in the December 15, 1898, issue of the *Oregonian* stated, "There promises to be in all the mining camps of the Greenhorn unusual activity next season. Before snow fell many quartz and placer locations were reported to have been made by prospectors." The account went on to observe that "mining men believe that all that country that lies between the middle and north forks of the John Day river will experience a mining boom before long." That same issue of the *Oregonian* also noted that between July and November of 1898, the Sumpter Valley Railway hauled 2,453,464 tons of ore and concentrates from Sumpter to Baker City.

ECONOMIC IMPACTS OF MINING

A measure of the economic impact of the gold flowing from the mines of eastern Oregon during the 1890s can be seen in the pages of a special edition of the Baker City *Morning Democrat* (May 20, 1898). Part illustrated souvenir and part chamber of commerce boastfulness, the special edition newspaper captured a snapshot of Baker City and Baker County just as the mining boom took off. It reported that the eastern Oregon gold fields contained 513 gold mines and claims in various stages of development. The mines employed about 1,200 miners, with placers yielding $1 million and quartz mines, $1.4 million during 1897 (historical dollar values). Five mines accounted for $1 million of the total lode production. Another measure of the steady increase in mining development over the decade was the addition of mill stamps at the mines. One commentator in the *Mining and Scientific Press* counted approximately 260 stamps in 1890, while another, in 1899, reported to the *Press* at least 441.[18]

Baker City, as the main supply point for the eastern Oregon gold fields, benefited greatly from all the mining activity. After describing the physical setting of the city, the writer of the *Souvenir Edition* noted:

> Baker City is not only the business and social center of the county, but is the center to which people flock, on trading or pleasure errands from hundreds of miles in almost every direction. Many of the business blocks of brick or stone cost from $15,000 to $50,000, and the Hotel Warshauer . . . is a modern three-story-brick,

> costing $70,000. During the past year there have been erected over 200 business structures, elegant residences and cottages at an outlay of $50,000, and the call for more room continues in spite of the utmost energy of every available bricklayer, mechanic and workman.[19]

By 1898, Baker City had electric, gas, and gravity water works, as well as a two-mile streetcar system. It included other aspects of a modern urban center of the period, such as a 50-bed hospital, $35,000 brick school building for 1,100 pupils, and an elegant opera house. The city also supported one daily and two weekly newspapers; a national bank; and industries, including machine works, sawmills, and foundries. It even had three cigar factories. The writer concluded his survey of the city's attributes by stating, "in Baker City are found blended to a useful degree the energy and conservatism of the pioneer and the culture and refinements of the Easterner. Yet the entire community of 7000 souls is leavened with that open-handed hospitality so apparent in every up-to-date and thrifty city in the great and growing West."[20] The writer, as it turned out, overstated Baker City's population; it did increase, however, from 2,504 in 1890 to 6,663 by 1900. During the decade of the 1890s, Baker County's population expanded from 6,764 to 15,597, an increase of 130.6 percent.

Other active mining communities in Baker and Grant counties showed similar growth during the 1890s. As Table 4.3 shows, Susanville (a 430.8

Table 4.3. Population of Select Eastern Oregon Mining Towns

Precinct	1870	1880	1890	1900	1910
Auburn	442	202	189	—	—
Canyon City	423	558	382	490	505
town	—	—	304	345	364
Susanville	103	150	52	276	280
Granite	448	200	249	836	148
town	—	—	—	245	89
Sumpter	—	261	91	1,857	684
town	—	—	—	1,809	643
Bourne	—	—	83*	592	220
town	—	—	—	—	77
Greenhorn	—	—	—	—	59
town	—	—	—	—	28
Baker City	312	1,258	2,604	6,663	6,742

**Cracker Creek precinct*

percent increase), Granite (a 235.7 percent increase), Bourne (a 613.3 percent increase), and especially Sumpter (a 1,940.7 percent increase), likewise experienced dramatic growth. Sumpter, surrounded by numerous rich mines, became a mushroom community after the narrow gauge Sumpter Valley Railway reached the town in the fall of 1898. Located thirty miles west of Baker City, Sumpter, according to the *Morning Democrat's* 1897 *Souvenir Edition*, had "well-stocked stores, good schools and hotels, a 50x100 opera house, and efficient water and electric systems. This terminal city has a growing future and enjoys a wide trade extending into the mining, stock and lumber sections." The *Souvenir Edition* also noted that the town had "a live weekly newspaper," the *Sumpter News*. The *Souvenir Edition* described Bourne, benefiting from the large production of the Eureka and Excelsior and North Pole mines, as connected to the outside world by daily stages and long-distance telephone and having "a good hotel and two well-stocked stores." Granite, in Grant County, also had numerous businesses, such as stores, two hotels, a school, five saloons, and a weekly newspaper. Granite and Sumpter, both confident of their staying power, sought and gained incorporation in 1900.[21]

A detailed look at the Susanville and Granite mining communities can reveal how profitable mining in eastern Oregon picked up during the 1890s. Both areas continued to sustain strong placer operations by white and Chinese miners, while developing and extending lode mining. Critically, outside capital provided the key to both types of mining. Investments in improved transportation, new types of equipment, and more sophisticated extraction processes assisted the effort. Still, it was remarkable that placers continued to produce gold from basically the same ground year after year between 1862 and 1900. Over time, though, the emphasis shifted to lode mining.

SUSANVILLE MINES

A closer examination of the Susanville District placers, especially along Elk Creek, demonstrated how the miners succeeded at this type of gold mining even after the best grounds had been worked over. Experts generally agreed that the Susanville District placers produced at least $600,000 in historical dollar values between 1864 and 1914. One even argued that the number might have been closer to $1 million, and none of the estimates apparently included the gold recovered by the Chinese, which had to be a substantial amount given their sizeable presence in the Susanville District in the 1870s and 1880s. The gold in this district was coarse, ranging from the size of mustard seed to a wheat grain. Occasionally, miners found nuggets worth up to $800 in historical dollar value.[22]

By the mid-1880s, the Elk Creek placer claims were said to be exhausted, although the Chinese apparently continued to work over the ground

Sloan and Haskell hydraulic placer mine on Elk Creek near Susanville, Oregon. Sluice boxes in foreground with steam derrick on the upper left. Courtesy of the Grant County Museum.

successfully. The *Grant County News*, for example, reported in its May 7, 1885, issue that twelve White and thirty Chinese were engaged in placer and lode mining at Elk Creek. According to an 1894 account in the *Oregonian*, while "some of the old-timers were still holding on and working, . . . the camp was nearly deserted and the town decaying." Into this situation stepped two experienced white miners, Horace Sloan and N. C. Haskell, who saw an opportunity. They reasoned that by controlling all the old claims and bringing in improved equipment and methods, they could make a handsome profit from the ancient grounds. Starting in the spring of 1883, they spent the next year or two getting control of about a mile of creek channel and then increased the water flow by constructing nine miles of ditch to divert water from another stream into their Elk Creek bed. The newspaper account went on to describe the partners' next steps: "They opened the ground from the mouth of the creek by a bedrock flume, added pipe and giant for hydraulic and a steam-hoisting engine and derrick for handling boulders, and began working over this drifted-out ground. . . . This ground has now been thoroughly opened for economical work by their method, and the past year the result has been most gratifying, and they have an assurance the future will amply reward them for

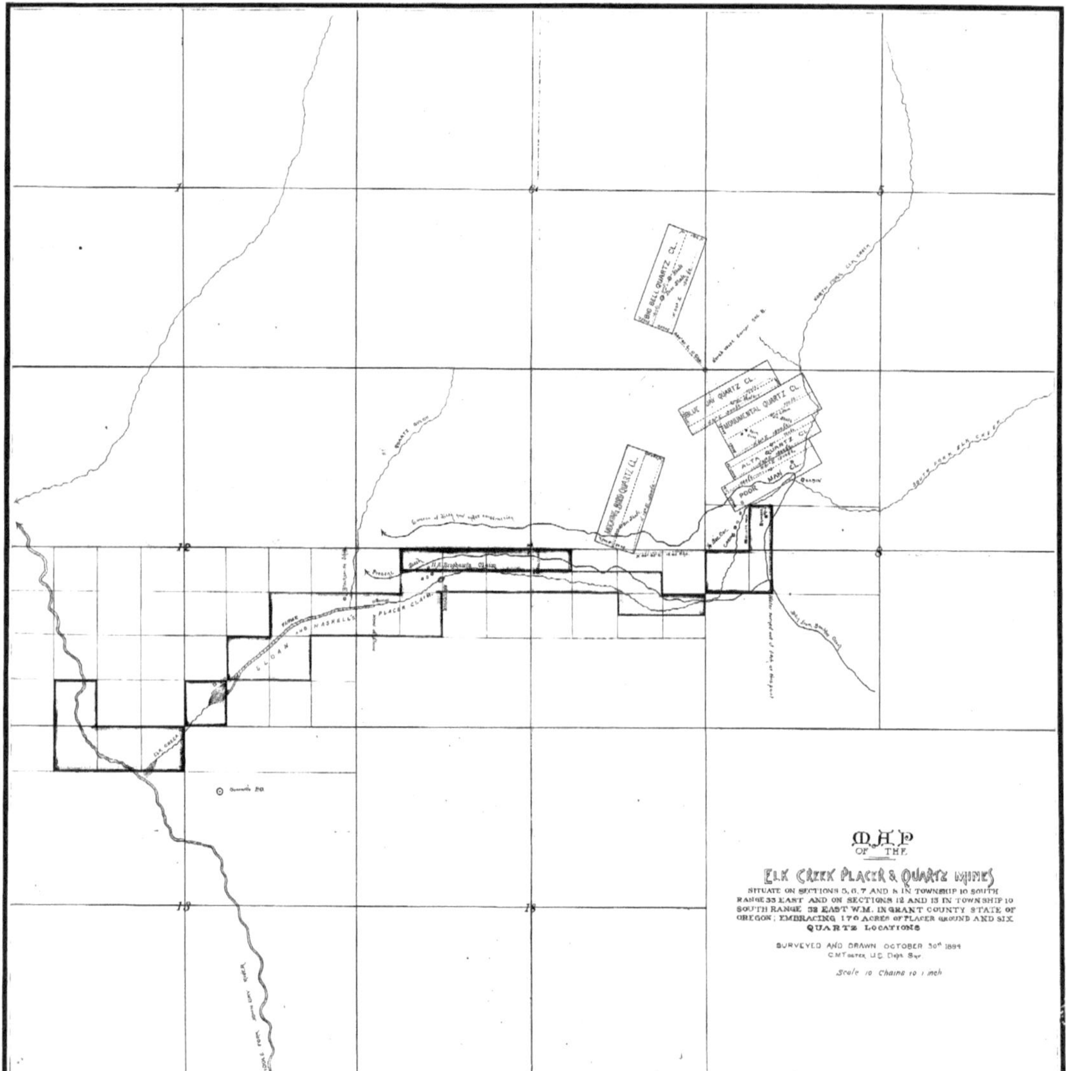

A General Land Office map of the Sloan and Haskell placer mine and quartz claims in 1894. It denotes several cabins, a steam derrick, a blacksmith shop, and the Susanville post office. Courtesy of the Oregon Department of Geology and Mineral Industries.

their faith in the enterprise."[23] In describing Sloan and Haskill's operation in 1890, the editor of the *Grant County News* noted rather romantically, that it was "now just about where old Susanville once stood—whose inhabitants have flown in the four quarters of the globe, leaving the place deserted and still, save for the continued roar made by the stream from the hydraulic nozzle as it plays against the gravel bank and liberates the precious metal we call gold from its long and dark confinement in the silent recesses of the earth."[24]

According to a later statement by N. C. Haskell, it took the partners ten years to complete all the preparatory work, during which they only managed to cover expenses. From 1893 to 1899, however, they turned a solid profit. During an annual season of six and one-half months and employing six to ten men, they extracted fine gold and nuggets worth about $70,000 in total. Their expenses came to $35,000, leaving a net profit of $35,000, "out of supposedly worked out, valueless, and abandoned placer claims. Since then we have disposed of the claims at a good price, and they are being operated profitably at this time [1903]."[25]

Operations did not always go smoothly for Sloan and Haskell. In November 1897, news accounts reported that a thirty-foot reservoir that they had just completed failed. Apparently, the floodgate was not strong enough to resist the increased pressure from the larger body of stored water. The dam "gave way, sending such a torrent of water, logs, stumps and other accumulated debris down the race that nothing could withstand it." The volume of water wiped out a second dam below the breached reservoir, flooding the "valley below, completely inundating the placer works." The flood took out "several hundred feet of flume . . . and the giant used in operating the water was knocked out of position, but only slightly damaged." In all, the Sloan and Haskell placer suffered about $1,000 in damages, but they quickly rebuilt the dams and flume.[26]

While Susanville District placers continued producing during the 1890s, lode claims received renewed attention. Miners turned to prospects located in early times and undertook new development work on them. Surveying what he called the once "flourishing" but now mostly deserted mining camp of Susanville, the editor of the *Grant County News* reported on April 17, 1890, that H. W. Sloan had several active lode prospects and an eight-stamp mill on Elk Creek as well as his hydraulic operation. In addition, other miners had discovered new prospects that seemed promising. In 1895, John Blake, a Susanville merchant for almost twenty-five years, purchased and began driving a tunnel on a quartz mine, the Gem of the Mountain, which had been discovered originally decades before. He patented the mine in 1898, after spending $1,490 on developing a tunnel 129 feet long and building a log house. In April 1898, news accounts touted a recent discovery called the Skyscraper Mine "as perhaps the richest mine on the coast."[27] It was said to have abundant ore, ranging in value from $50 to $350 per ton. W. P. Mount, who had bonded the Skyscraper, stated his intention to fully develop the mine. Over time, however, the Skyscraper never lived up to its promoters' early billing. The excitement generated by the prospecting and early development activity in the Susanville District reached a fever pitch during the summer of 1898. The editor of the *Long Creek Eagle* declared that "Susanville will come to the front as the most flourishing mining camp on the coast."[28] Other claims

Table 4.4. Principal Lode Mines of Susanville Mining District

Mines	Patent	Peak Production Years	Estimated Production	Year Located
Badger	Yes	1870–1874, 1899–1905	$500,000	1869, 1878, 1897
Gem	Yes	1898–1910, 1916	Small	1869
Bull of the Woods	Yes	1897–1905	Small	1870s
Poorman	Yes	1901–	Small	
Princess	No		Small	1879
Side Issue (Black Hawk)	Yes	1903–1907	Small	1903
Stockton	Yes		Small	

being vigorously investigated in the late 1890s included the Poorman, Blue Jay, Rambler, and Bull of the Woods.[29]

The one Susanville District lode mine that did satisfy its owners' golden dreams was the Badger group of seven patented claims. Originally discovered around 1869 and then prospected again around 1878, the claim was purchased by Susanville merchant and miner John Hughes in 1883. He continued minor development to a depth of 90 feet and extended drifts on the vein 265 feet. His exploratory efforts through 1897 indicated that he had ore valued at $35 to $140 per ton—enough to justify a major investment in the property. Lacking the finances to personally continue development, Hughes, in July 1898, sold the mine and other prospects to a San Francisco syndicate for $12,000 plus fifteen percent royalties on ore shipped while payments were pending. In 1899, the purchasers reorganized the mining operation as the Badger Gold Mining and Milling Company.[30]

The new company determined that the mine had high-grade ore above the 250-foot level of the inclined shaft, running from $50 to $300 per ton—better than initially indicated. Based on the early profitable returns from the smelter, the Badger Company began a full-scale buildout of the operation. They constructed an up-to-date ten-stamp gold mill, cook and boarding house, company office, bunkhouses, ore houses, and a hoisting plant with a 40-horsepower engine. While this activity got underway during the fall of 1898, the company mined and shipped out almost 50 tons of ore. Mining and further exploratory work continued until miners had driven a 900-foot shaft, a 1,600-foot crosscut adit (a tunnel crosswise to the vein) on the 500-foot level, and several hundred feet of drifts (an adit following the vein). Unfortunately, at the lower levels of the mine, the ore contained higher percentages of sulfide. As will be discussed in the next chapter, processing these lower-grade ores to extract the gold content was costly, and the high transportation cost for getting the ores to the smelter in Tacoma, Washington, made it unprofitable to continue mining after November 1905. The Badger Company also became entangled in a lawsuit with an adjoining mine claim, which reinforced the decision to shut down operations. During the years it functioned, the Badger Mine employed between 30 and 60 men.

The nature of the Badger ores was not a unique situation in the Susanville District. Other lode or quartz prospects in the vicinity of Elk Creek soon revealed a similar limiting factor that raised questions about the long-term viability of the Susanville mines. One mining commentator succinctly stated the ore challenges facing Susanville miners:

> The best veins of quartz so far are found between a granite and porphyry gangue of rock, mostly in the latter. The ledges vary from 4 to 20 feet in width, and have the appearance of permanent, true fissure veins. They carry free gold and some silver on the surface, and an assay of from $4 to $90 on the surface croppings, but as soon as water or depth is attained they change to base and refractory ores, carrying a percentage of arsenical copper, antimony and ruby sliver, together with gold sulphurets, and average assays run from $3 to $20 per ton.[31]

The expense of treating the lower-grade ores with the technology of the times, especially if the veins were not extensive, made most of the prospects ultimately unprofitable.

GRANITE MINES

During the 1890s, precious metal mining in the Granite District followed the same trajectory as that in the Susanville District. Placer mining continued apace, while lode mining steadily increased. The Granite District lode mines, however, outpaced the Susanville District lode mines in productivity. The most prominent and long-lasting placers included the Tabor placer on a Clear Creek bar four miles below the town of Granite, the Klopp and the Thornburg placers on the North Fork of the John Day River, and the Crane's Flat placer six miles northwest of Granite. The application of hydraulic mining technology during the 1890s increased the recovery from old claims and led to the opening of new ground. By the early 1900s, miners began dredging to reach more deeply deposited gold found on the bedrock of streams. Although Waldmar Lindgren stated in his authoritative book *The Gold Belt of the Blue Mountains of Oregon* that he did not think the Granite placers were especially rich, other experts credited the placers with yielding about $2 million in historical dollar values through 1914.[32]

While the quantity of gold and silver from the Granite District lode mines exceeded that of the Susanville District, it still was not as great as that from the Cracker Creek District to the northeast. Miners considered the Granite

Granite, Oregon, in 1894, with Main Street in the foreground and Chinatown buildings located on the upper right near Granite Creek. Courtesy of the Baker County Library.

Mining District to be the southwestern extension of the so-called "Mother Lode of the Blue Mountains," located in the Cracker Creek and Cable Cove mining districts. This belt was thought to run through eastern Oregon in a band approximately 100 miles long and thirty-five miles wide. Miners speculated that the lower elevation placer gold had washed down the area's many small streams from veins located in the higher elevations of the divide between Cable Cove and Granite districts. Accordingly, prospectors logically explored lode possibilities higher in the mountains. By the mid-1880s, a number of lode prospects had been discovered, but only limited development occurred initially. The Monumental Mine operated only sporadically during the 1890s because the price of silver was too low to earn a profit. Other, small mines in the Granite District included the Ajax, Continental, Ibex, and Magnolia. The location of and work at these mines took off in 1898/1899, and the town of Granite boomed as a result. In the long run, most of the lode gold in the Granite District came from three properties: La Belleview, Buffalo, and Cougar-Independence. Altogether, the lode production of the district amounted to about $2.2 million in historical dollar values from approximately ninety mines and prospects.

Fred Cabell, who began his mining career in the Elk Creek area in 1864, discovered one of the earliest quartz claims (1869) there. Cabell, however, soon began prospecting in the Granite and neighboring districts. In 1876, he discovered a prospect near the head of Onion Creek, naming it the La Belleview Mine. The mine, which ultimately included six patented claims, contained both gold and silver; however, the early work using an arrastra yielded limited results. In 1890, Salt Lake City mining investors bought the property and patented the mine in 1893. Under the new ownership, miners dug 6,000 feet of underground tunnels from four adits over a vertical range of 600 feet. The mine sat at an elevation of almost 7,000 feet. The total production of precious metals through 1911 came to $200,000 in historical dollar values. The concentrates shipped to the smelter for processing averaged 1.2 ounces of gold and fifty-five ounces of silver per ton.

Discovered in late 1899, the Cougar-Independence mine, at an elevation of 5,000 feet, at first had limited success. The mine consisted of two parallel veins about a mile apart: the Independence and the Cougar. In time, however, the Independence vein proved less valuable. Both mines had the same ownership and used the same mill to process the ore, which was difficult to separate from the base material. Development work at the Cougar resulted in about 5,000 feet of drifts from four adits, while effort at the Independence consisted of an upper tunnel 250 feet long and a lower level of 1,020 feet with a 1,200-foot crosscut. The owners did not attain real success until the 1938–1942 time period, when they installed a new mill and concentration equipment to better handle the recalcitrant ore. The two mines ultimately produced gold and silver worth $750,000 in historical and modern dollar values.

With a production history stretching back to the mid-1880s and continuing into the 1980s, the Buffalo Mine has the distinction of being one of the longest operating lode mines in the Blue Mountain gold region. In 1880, six brothers by the name of Beagle located the three lode claims initially encompassed by the Buffalo-Monitor mine, located at an elevation of about 6,000 feet. Ultimately, the Buffalo-Monitor mine included thirteen claims, four of which were patented. Geologists speculated that the quartz vein of the Buffalo mine was an extension of the Cougar and Magnolia vein system. The gold found on the surface was free-milling and processed with an arrastra. Once the miners exhausted the free-milling ore, more complex milling processes became necessary. Transportation costs to the Tacoma smelter for such complex methods became too costly to sustain mining. New owners took over in 1903 and had better success. While no production records exist prior to 1903, the period between that year and 1909 resulted in $75,000 in historical dollar values. Mining continued under different owners between 1919 and the 1980s, with a recorded production value of $900,000 in modern dollar values up to 1958. Over time, miners expanded the underground work to include 10,000 feet of drifts and crosscuts covering five parallel veins. There are four adit levels, the lowest at 600 feet below the surface.

The Red Boy Mine, in the Granit Distric, Grant County Oregon.
Owned by Godfrey and Tabor. This Picture was taken about 1896
The approxmate production of the Mine in Gold, One Million
dollars.

The famous Red Boy Mine near Granite, Oregon, ca. 1905. One of the most productive and best-known mines in eastern Oregon. Courtesy of the Grant County Museum.

Another famous mine, five miles southwest of Granite—the Red Boy—also came into prominence during the 1890s. Originally considered part of the Granite Mining district, it later became listed as a separate district, surrounded by a cluster of other mines that seemed extensions of the Red Boy and two other parallel veins. In time, though, mining experts included the Red Boy with mines comprising the Greenhorn District. Although discovered in the early 1880s, initial development failed, and its owners stopped working the mine for a time. As Lindgren, a leading mining expert, put it, "The Red Boy for many years was worked intermittently and with indifferent results."[33] The situation changed in 1893 when A. G. Tabor, E. J. Godfrey, and Clark Taber took over the property and installed an improved Crawford mill (a newly patented machine for grinding ore to a fine consistency in preparation for concentration or amalgamation). The mine soon paid well, and the new owners continued with further improvements in 1898, including a new twenty-stamp mill, cyanide plant, two-mile-long covered ditch providing waterpower to four Pelton wheels, and an expensive hoist system. To support mining operations,

Table 4.5. Principal Lode Mines of Granite Mining District

Mine	Patent	Peak Production Years	Estimated Production	Year Located
Ajax	No	1905–1906	$40,000	1898
Buffalo-Monitor	Yes	1880–1903, 1903–1965	$900,000 1878/1888	
Continental	No	1880–1926	$50,000	1880
Cougar-Independence	Yes	1900–1911, 1922–1950	$750,000	1898
Ibex	Yes	1899–1901	Small Yield	1888/1901
La Belleview	Yes	1878–1892, 1927–1929, 1939–1941	$500,000 ($200,000 b/w 1878–1911)	1878–1887
Magnolia	No	1899–1904	Small Yield	1899
Monumental	No	1875–1895	$100,000	1870
New York	No	1899–1906	?	1901–1909

the owners erected a boarding house, shops, and other outbuildings. In its *Souvenir Edition* of May 20, 1898, the *Morning Democrat*, claimed that the Red Boy "and the adjacent claims have attracted the attention of some of the largest mining operators in the country, but all propositions to purchase have been rejected by the owners." Site developments underground consisted of 5,000 feet of tunnel from three adits and a 300-foot shaft. The mine typically employed about fifty workers. In time, the rich upper ledges of ore gave out, and lower-grade base ores proved too expensive to treat. Excessive water in the lower levels also increased expenses. Between 1890 and 1914, the Red Boy mine produced about $1 million in historical dollar values.[34]

RANCHING ALONG THE MIDDLE FORK OF THE JOHN DAY RIVER

Mining was not the only economic activity in the Susanville area in the 1890s. Between 1888 and 1902, ten settlers patented homesteads totaling 1,600 acres, while two others made cash purchases of 160 acres each in the Susanville precinct. While the area was mostly mountainous and heavily timbered, these settlers took up fertile bottomlands along approximately nine miles of the Middle Fork of the John Day River in sections of T10S, R32E and 33E. Such well-watered meadow lands produced abundant crops of hay and alfalfa that fed livestock and supported small orchards and vegetable patches. As early as 1885, the *West Shore* magazine, in touting the attractions of Grant County, noted the availability of good farmland along the Middle Fork. As a writer to the *Oregonian* explained in 1898, while "Grant is a mountainous county [,] it is liberally interspersed with beautiful and fertile valleys. The elevated portions furnish an abundance of rich and nutritious grass for summer range, while the valleys produce sufficient hay to care for the many herds during the winter season."[35] The next chapter will discuss in detail the demographic characteristics of the Susanville homesteaders.[36]

MINING DURING THE ECONOMIC DEPRESSION OF THE 1890s

It should be noted that the eastern Oregon gold boom of the 1890s occurred during an otherwise difficult economic period for Oregon and the nation. From 1893 to 1897, the United States suffered its worst economic depression to that time. Across the nation, 500 banks failed, and 16,000 businesses went bankrupt, including major railroads and manufacturing concerns. The result was widespread unemployment and misery. Although probably an exaggeration, the governor of Oregon estimated that two-thirds of the state's workers were without jobs at the height of the depression. Agricultural prices plunged for a time, and farm mortgage foreclosures increased in the Pacific Northwest. In 1894, the price of wheat, Oregon's most valuable agricultural product, plunged to $0.43 per bushel; in 1891, it had been $0.88. In Grant County, the price of wheat fell as low as $0.26 a bushel. Statewide, the value of farms devoted to the production of wheat decreased from more than $11 million in 1891 to under $4.5 million in 1894. Wool, another major agricultural item in Oregon, also took a hit, as the removal of a tariff in the early 1890s allowed foreign producers to flood the United States with cheap wool at a very inopportune time for those in the sheep business.[37]

Amidst this gloomy backdrop, the uptick in eastern Oregon mining offered some relief. It brought in outside investment and put new money into circulation, either as wages or profits. The anonymous author of the *An Illustrated History* . . . wrote in 1902, "The next year [1893], however, brought panic and distress to the country generally, brought to Grant county its share of depression." But he went on to state, "Then, too, the depression and consequent cheap labor rendered it possible to work mines which would not pay the scale of wages obtaining in prosperous season, so that a means of obtaining a livelihood was thus furnished to a portion of the population." That same author, in writing about the depression's effects on Baker County, noted: "The years 1894 and 1895 were not productive of much advancement in any line," however, "a mining community, especially a gold producing community, seldom feels financial depression as do manufacturing and agricultural sections."[38]

The increase in Baker County's assessed valuation over the decade of the 1890s would seem to bear out the notion that gold mining alleviated the worst consequences of the Depression of 1893 on eastern Oregon and, by extension, on the state to some extent. In round numbers, Baker County's valuation grew from $2.0 million in 1890 to $2.9 million in 1900, an increase of 42.8 percent. For context, Multnomah County, the home of Portland, actually experienced a decrease from $39.6 million in 1890 to $32.6 million in 1900, a drop of 18 percent in valuation. Umatilla, an eastern Oregon county dependent on wheat and sheep, saw a small increase in its assessed valuation, going from $4.3 million to $5.6 million (an increase of 27 percent) during the same time frame. The assessed valuation of Oregon rose a modest 3.3 percent: $114 million in 1890 to $117.8 million in 1900.[39]

After a decade of very active mining investment and development, accompanied by considerable press hype, the total output of Oregon's mines remained steady except for slight jumps in 1891 and 1894 (see Table 4.2). In fact, Oregon's entire production of gold and silver ranked it tenth in both 1890 and 1899 among the precious metal producing states and territories west of the Mississippi River. Many unpredictable factors could affect mining operations. Weather, for example, directly affected the success of placer mining. The amount of rainwater and snowpack moisture determined the length of time that placer mining could be carried out each year. Lode mining had its own difficulties. Success depended on the abilities of managers, how well machinery and chemical processing functioned, and the richness of the ore. Ultimately, only expensive, exploratory tunnelling could determine the quality and quantity of a gold vein. As new mines opened with great expectations, old ones would shut down after running out of paying ore. Still, between 1877 and 1899, Oregon's placer and lode mines produced 1,342,525 ounces of gold worth $27,749,997. The question on everyone's mind was what the new century would bring to eastern Oregon's precious metal industry.[40]

Prosperous Baker City, Oregon, ca. 1910. View of the stone and brick buildings on Main Street, looking north. Courtesy of the Baker County Library.

Chapter Five
The Gold Fields of Eastern Oregon

The new century began with great expectations of growth and prosperity for Oregon. Nowhere were such hopes so ardently held as in the mining region of eastern Oregon. Headlines in the *Oregonian*'s New Year's annual edition for 1900 trumpeted, "Mines of Great Wealth—Experimental Stage Has Passed and the Era of Development Has Begun." The newspaper noted that "Substantial as was the achievement of 1899, it will be left far behind by the development of 1900. . . . Under the stimulus of modern machinery in extracting values, the Eastern Oregon belt will this year advance to its proper station as one of the world's great gold producing regions."[1]

The newspaper writer speculated that by 1901, the annual output of eastern Oregon mines would more than double. At the time, he reported that the eastern Oregon gold belt had nine highly productive mines with mills containing 195 stamps, producing an annual output worth $1.45 million in historical dollars. To that figure, he added $175,000 from irregular producers and another $600,000 from placers. He then estimated that the new development underway at six existing mines, coupled with new machinery at eight new mines projected to add mills with 185 stamps, would be capable of producing another $2.65 million worth of gold. All this mining activity would take place in three mineral zones stretching from the Seven Devils region of western Idaho along the Snake River to Susanville on the Middle Fork of the John Day River in Grant County—a distance of approximately 160 miles in length and fifty miles in width.[2]

The same issue of the *Oregonian* contained a lengthy sidebar detailing the recent capital investment in the eastern Oregon gold belt, particularly around the town of Sumpter. Especially notable were the number of out-of-state and foreign capitalists pouring money into the mines. The investments brought in new machinery to reach the gold buried at greater depths than operations had yet attained. By 1900, development at the most productive mines had reached no more than 1,000 feet below the apex or outcrop of the surface ledge of minerals. The spectacular recent investments included the Bonanza Mine bought by Philadelphia investors for $750,000 and the Columbia Mine and an adjoining claim for $130,000 by a lumber merchant from Minneapolis. The author of this column went on to list numerous other purchases and equipment investments in the $50,000 to $100,000 price range.[3]

A correspondent from Baker City, writing in the same *Oregonian* annual edition, chided Portland capitalists for not investing in eastern Oregon mines and thereby missing a "golden opportunity":

> Mining men all through this section say Portland is asleep. Is this true: Are the business men of Portland getting hold of property in Eastern Oregon? If not, why not? Is it because you are afraid to invest? Do you think the values are not here? Then send out your experts to investigate. Do you think capitalists from all parts of the United States and British Columbia would have invested no less than $5,000,000 in property upon the mere supposition of its worth?[4]

The writer went on to assert that Portland would gain more than mere riches from investing in eastern Oregon gold mines:

> By locating and purchasing property in the mines, Portland will not only share with the rest of the world in the rich harvest of gold from the ore, but there will be established between her people and the miners a feeling of kinship which will readjust the trade relations now so sadly strained. Generations will be born, grow old and pass away before the mills will cease to separate the gold from the rock in Eastern Oregon. No one more than the native and adopted sons of Portland is entitled to the benefits.[5]

At the very least, the writer thought that Portland interests should build a smelter, which could process the mineral wealth of not only eastern Oregon but also that of the whole Pacific Northwest. As he summed up his argument, "The benefit of the right kind of smelter, with the spirit of giving in it, will mean more to Portland and the state than it has entered the mind of one to conceive."[6] In the same vein, the editor of the *Pacific Miner*, noting the money to be made in business dealings with the thriving mining communities of eastern Oregon, asked "Why is it . . . that the people of Portland cannot see the vast importance of the mining industry of the Eastern section of the state and help boost the cause of those who are endeavoring to give it desired prominence?"[7]

A promotional pamphlet for the Oregon Railroad and Navigation Company published in 1900, echoed the excitement and opportunity expressed by the *Oregonian*:

> The gold-ribbed mountains [of eastern Oregon] have thrilled with new life. Thousands of prospectors have flocked in from every part of the globe, and the grand army of actual mining works has more than doubled, probably trebled, possibly quadrupled. . . . The whole far stretching mineralized region has been dotted with new and flourishing camps—one or two which, at least, bid fair to rank among the wonders in mining annals.

The writer concluded his panegyric by claiming that "all over this gold-strewn wonderland new discoveries of riches follow each other like cards from a practiced dealer's hands."[8]

Golden opportunity seemed available everywhere one looked in turn-of-the-century Baker and Grant counties. As the Baker City Chamber of Commerce averred in one promotional piece, "today it is safe to say that there are 5,000 locations on the gold ledges around Baker City. Of this number, at least 20 per cent will develop into good paying mines. Some have already produced a million, some hundreds of thousands of dollars, and only a beginning has been made—a mere scratch on the surface of a vast mineral territory the extent and real value of which cannot now be estimated."[9] The explosive growth of the town of Sumpter reflected the lure of such overheated promotion. In 1897, at the beginning of the eastern Oregon gold boom, the town had scarcely 300 inhabitants. According to the 1900 federal census, Sumpter had mushroomed to almost 1,900 residents (Table 4.4, previous chapter). It sat at the opening to the richest gold deposits in eastern Oregon. The Cracker Creek mines were a mere six or seven miles distant; Granite District mines, fourteen; the Bonanza District, twelve; and Susanville mines, sixty miles. As a writer in the *Pacific Miner* put it, Sumpter, "that thriving mining town of Eastern Oregon can claim that within an area of a few miles from its center there are a greater number of paying mines than in any other section of the Pacific Northwest." He went on to wax poetic: "No fairy's wand is needed, no magic employed to convince one of the fabulous richness of the mines of the Sumpter region."[10]

Almost overnight, Sumpter, incorporated in 1900, boasted such modern amenities as electric lights and piped water. At the turn of the twentieth century, its business center consisted of two banks, six hotels, sixteen saloons, two hardware stores, four general merchandise stores, drug stores, three tailor shops, warehouses and repair shops, livery stables, a large sawmill, lumber yards, a brewery, and two newspapers. A fancy opera house, dance hall, and a dozen fraternal orders supplied entertainment, and schools, churches, and a hospital provided needed services. In addition, Sumpter, according to the 1900 federal census, had a thriving "Red Light" district. The census taker listed fifteen female prostitutes and eight barmaids. Professionals present in the town included the usual lawyers and doctors and such specialties necessary for a mining community: mining engineers, mineral assayers, and real estate and mining stock promoters. Sumpter also served as a major trans-shipping point for mining equipment, with the *Sumpter Miner* reporting on November 14, 1900, that five million pounds of mining machinery had been brought to the city during the year by the Sumpter Valley Railway for final transport by horse teams and wagons to the various mines in the surrounding mining districts. In 1900, the Sumpter Valley Railway carried 33,934 passengers to and from Sumpter. By 1902, at the height of the mining surge, the editor of the *Sumpter Miner* boasted,

> "the City of Sumpter is an up to date progressive camp of three thousand inhabitants, with every requisite of a flourishing city of that size. We have all modern conveniences, electric lights,

A view of boomtown Sumpter, ca.1905, looking east with schoolhouse and hospital in upper center of photograph. Courtesy of the Baker County Library.

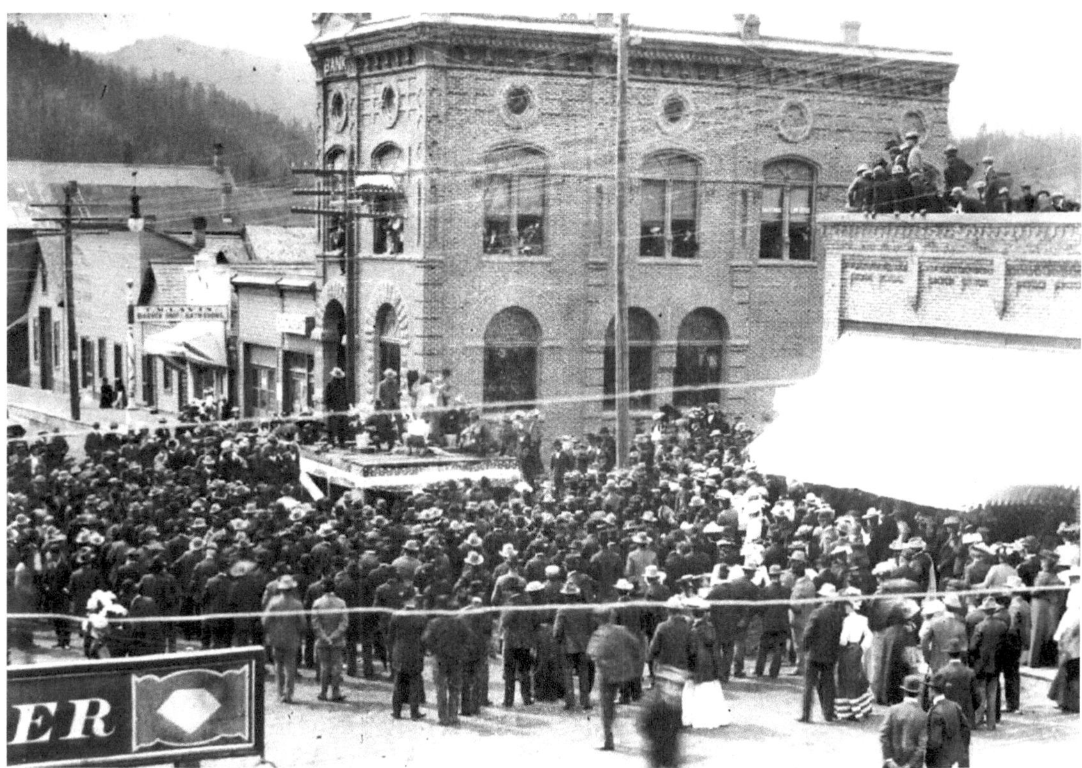

The Bank of Sumpter with a crowd watching a rock drilling contest. Courtesy of the Baker County Library.

> telephone and telegraph connections, two solid banks, fine climate, pure water, good schools, churches, and we wish to emphasize the fact that statistics show our section to be the most law abiding mining community in the world.[11]

The booming of golden Sumpter typified, as historian Marks described, "the urban communities that exploded helter-skelter from the mining camps at or near strikes."[12]

Susanville also grew quickly in the first years of the twentieth century. By 1901, it had split into two communities. "Old Susanville," soon to be renamed Galena, sat below the mouth of Elk Creek on the Middle Fork of the John Day River. It had a hotel, a meat market, a blacksmith shop, two saloons, a sawmill, a laundry/barber shop, and two general stores—one run by a Caucasian man and the other by a Chinese man. The old post office moved a mile up Elk Creek to the "New Susanville" next to the Badger and other nearby mines. The Badger Mine Company opened its own general store and boarding house, and additional businesses followed at this location as well. In response to local demand, the federal government established a new post office at Galena late in 1901.[13]

In 1900, Grant County also experienced a feverish search for gold. The January 25, 1901, edition of *Blue Mountain Eagle* reported that during the preceding year, the county clerk recorded 1,650 mining locations, "principally on quartz property." Of course, as the newspaper noted, "The number of

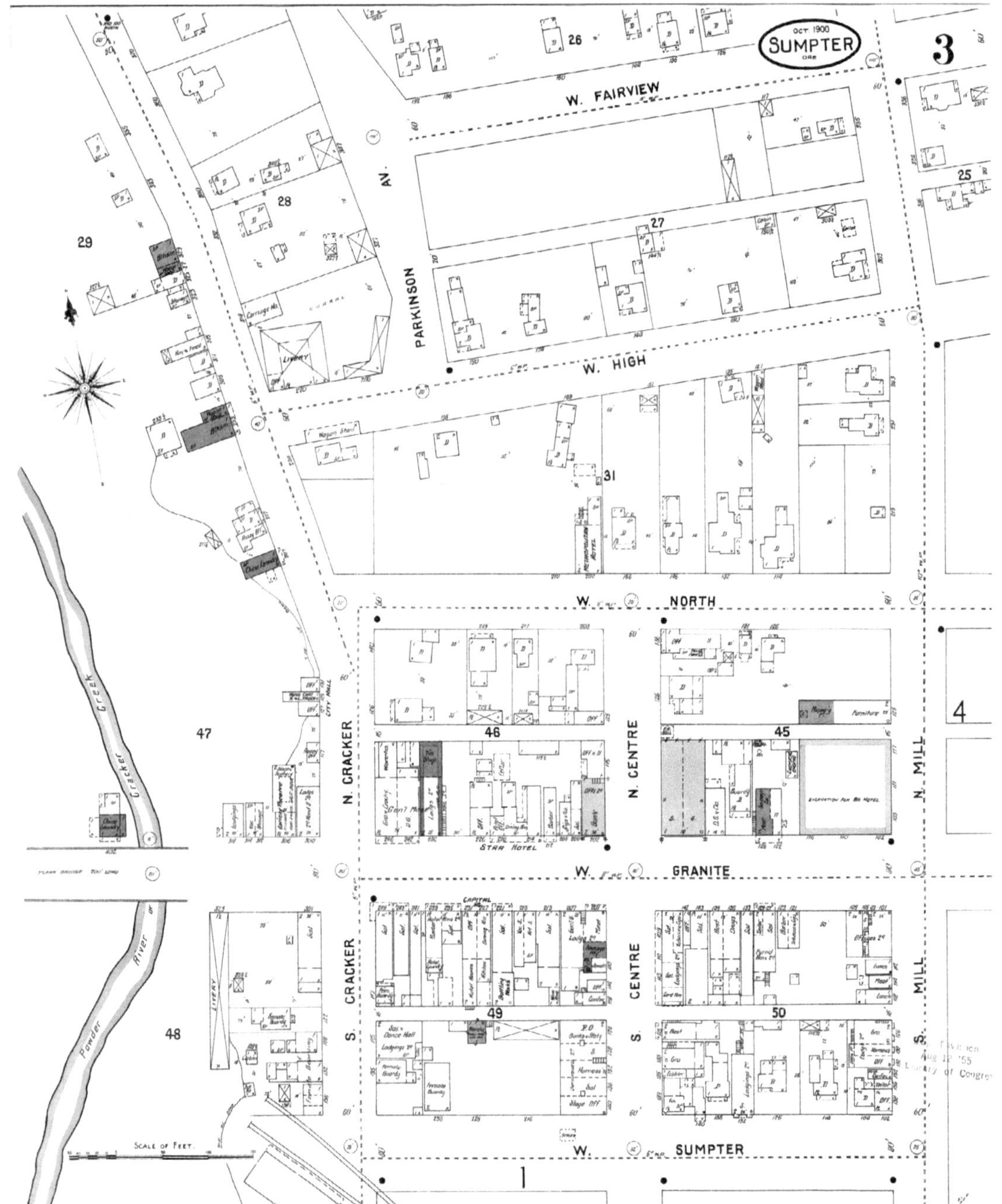

Sanborn Insurance Map of Sumpter, 1900. The red-light district is shown in the lower left of the map. Courtesy of the OHS Research Library.

these locations that will develop into good paying mines remains to be seen." The most active prospecting occurred in the Greenhorn, Susanville, and Quartzburg mining districts of the county. After summarizing all the mining activity at Susanville in 1900, a correspondent to the *Sumpter Miner* declared, "The Susanville District has ore, water and timber and I predict within the next year results from that section will astonish the world."[14]

KEY MINING DISTRICTS

While the Blue Mountains of eastern Oregon had thirty officially recognized mining districts during the period from 1890 to 1910, only three—Rock Creek, Cracker Creek, and Greenhorn—contained the most productive lode mines. Other districts, such as Granite, Susanville, Sanger, and Sumpter, had placer and lode mines exhibiting intermittent or limited production during the two decades under study. As Table 5.1 indicates, the Cracker Creek District contained some of the most valuable mines. The following discussion begins with the Rock Creek District at the north end of the eastern Oregon gold belt.

The Rock Creek District sat in the Rock Creek and upper Pine Creek drainages on the eastern slope of the Elkhorn Range. This mountainous area was about fourteen miles west of Baker City and had an elevation that ranged from 5,500 feet to 8,500 feet. The gold veins, which extended northeast to southeast, appeared in the contact points between granodiorite and argillite country rocks. Although the district had thirty mines and prospects, the principal ones consisted of Baisley-Elkhorn and Highland-Maxwell. Located on the same vein, those two yielded at least $1.5 million (Table 5.1). Prospectors had located the two mines in the late 1880s, and they remained active until about 1910, with intermittent production continuing up to 1940. Although both have incomplete records, Baisley-Elkhorn represented the richer of the two properties. Both mines had twenty-stamp mills and extensive underground works: 10,000 feet of development at the Baisley-Elkhorn and 15,000 feet at the Highland-Maxwell.[15]

The Cracker Creek District extended 10 miles northeast to southeast in the Elkhorn Range and included the drainages of Cracker and McCully creeks. It straddled the divide that separated the Powder and North Fork of the John Day rivers. The area was located six miles directly north of Sumpter, with an elevation ranging from 5,500 feet to 8,000 feet. The gold deposits resided in complex formations of quartz-rich veins embedded in fractured zones, consisting of argillite and granodiorite intrusive rocks. In 1901, Waldemar Lindgren, a highly respected government geologist, noted the interconnection of the Rock Creek and Cracker Creek districts, describing them as "practically one continuous vein system, beginning at the Baisley-Elkhorn mine and continuing across to Cracker Creek, with a strike varying from east-west to northwest-southwest, and some of the most important mines in eastern Oregon are comprised in it. . . . It is easily the most strongly defined and persistent vein in the Blue Mountains." The settlement of Bourne served as the supply point for the mines of the Cracker Creek District.[16]

While the Cracker Creek District had more than fifty mines and prospects at its peak development, five mines stood out as particularly rich (Table 5.1). The five mines stretched slightly over five miles along a vein known as the North Pole–Columbia lode, but the most productive portion of the lode extended about 12,000 feet. Development work for the five mines covered almost 100,000 feet on several levels, reaching a depth of 1,000 feet at the Columbia Mine. Prospectors located the North Pole–Columbia vein in the 1880s, and investors patented and developed—from north to south—the North Pole, Eureka and Excelsior (E&E), Taber Fraction, Columbia, and Golconda mines from the 1890s to about 1908. The investment firm Baring Brothers of London owned the North Pole Mine, Portland lawyer and politician Senator Jonathan Bourne and associates held the E&E Mine, Edward Backus and associates of Minneapolis controlled the Columbia Mine, and

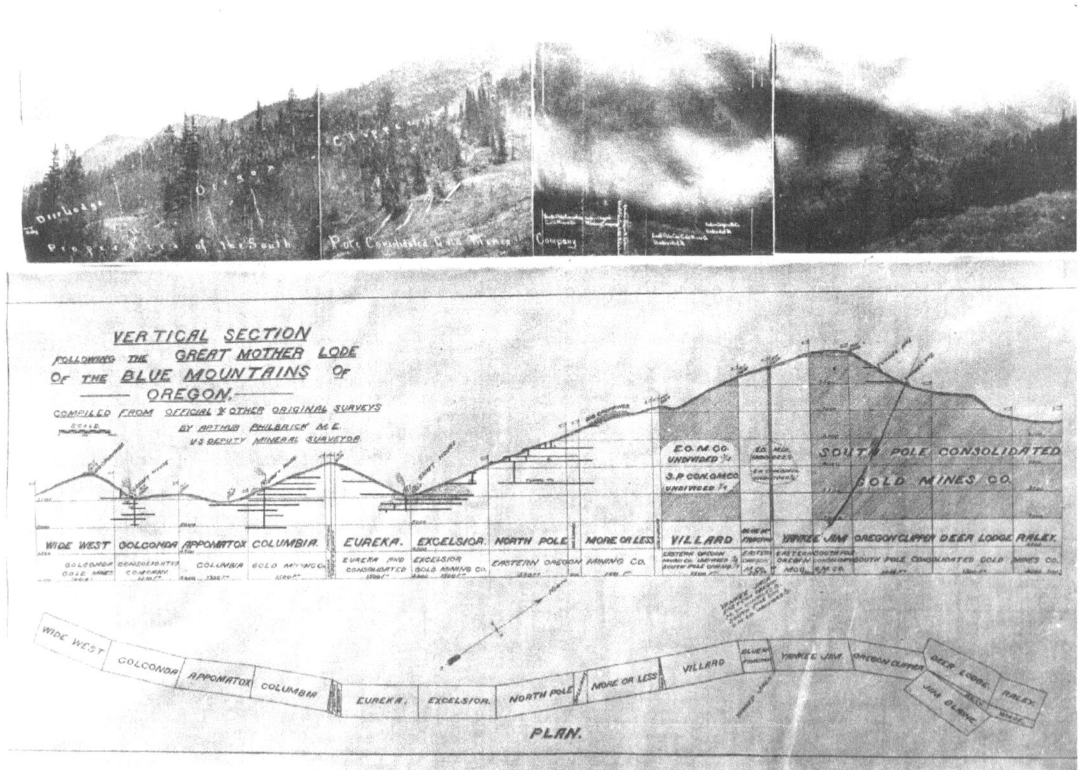

A vertical and horizontal view of the Mother Lode of the Cracker Creek Mining District. Courtesy of the Baker County Library.

A view of Senator Jonathan Bourne's Eureka and Excelsior Mine mill and other buildings, ca. 1900. Note the man standing at the door. Courtesy of the Baker County Library.

Portland newspaper publisher C. S. Jackson had the Golconda Mine. Legal disputes kept the E&E Mine idle between 1898 and 1903 while the others attained some of their best results in those years. The Golconda, Columbia, and E&E were equipped with twenty-stamp mills, while the North Pole had a thirty-stamp mill. All the mines had extensive chemical processing plants. The daily production from each of these mines ranged between fifty and sixty-five tons. Typically, operations ran year-round, employing between fifty and seventy-five men each. Incomplete records indicated the total production from the district at almost $9 million.

Because the milling and concentration processes used at the time in the Cracker Creek mines only recovered about 67 percent of the ore value, losses in the tailings were about $4 million. To recover the lost value and continue profitable mining would have required significant investments in the newly emerging improvements in mining technology. Writing in 1916, about the most valuable Cracker Creek mines, H. M. Parks of the Oregon Bureau of Mines and Geology observed that recent developments in cyaniding complex ores and floatation processes would allow recovery of 90 percent of ore values, but owners and managers were reluctant to make the necessary improvements because it "requires extensive alterations in and additions to the present mill[s], which would involve the expenditure of considerable sums and would absorb their dividends for some time." Parks further noted, "a proper consolidation of these properties is an economic necessity for most of them and would be highly beneficial to all. Attempts to consolidate the leading properties have been made by some of the owners, as well as by outside interests, but for one cause or another have been unsuccessful."[17] He then went on to describe the chief difficulties that stood in the way of uniting the mining properties under a single ownership:

> The owners of these properties are practically all men of affairs actively engaged in banking, publishing, politics and industry. Their multiplicity of interests causes all but one property to be kept idle. Their chief interest lies not in mining, and their experiences in it were for most of them secured during a period when close valuations upon mining properties were much less common than now, and the experienced engineer, metallurgist and mining geologist, now so prominent in the operation of successful mining companies throughout the world, had then but a small part in operations. A failure to fully realize that experienced technical men can solve their problems of mining and milling keeps some of them from operating their properties themselves, and the experience of some of them during the time when the element of adventure existed to a greater degree than now causes some to over-value their property when considering its sale.

Parks also cautioned that mines left idle for any length of time did not fare well:

> Mine openings are not permanent ones. They have a considerable annual depreciation, and most of them, if left idle for a decade or two, will become nearly a total loss. The only factors which increase the value of known bodies of ore are reduction in the cost of and an increase in the percentage of extraction. But the idle property, with its rapid depreciation in the value of mine openings and loss of returns upon capital invested during the period of idleness, will in [the] future vastly exceed any gain from improvement in processes.

The Greenhorn District, southwest of the Cracker Creek mines, included portions of the drainage of the Upper Burnt River and the Middle Fork of the John Day River and extended on the east as far as the Geiser Creek area. The

Main Street, Greenhorn, Oregon, ca. 1910. Courtesy of the Baker County Library.

district sprawled across both Baker and Grant counties and, by one estimate, had as many as eighty-four mines and prospects. Elevations reach as high as 8,200 feet (Vinegar Hill). The country rock consisted mostly of granodiorite, argillite, and greenstone with gold deposits in quartz veins. Some deposits also had a high silver content. The chief mines included Bonanza, Red Boy, Pyx, and Ben Harrison. Writing in 1900, Lindgren expressed skepticism about the prospects of the Greenhorn District, stating that "few of these deposits can be ranked as mines, inaccessibility and expense of treatment having retarded this section very much." Early prospecting had led to the founding of the settlement called Robinsonville in 1865, but it burned down in 1898. A second town, Greenhorn, sprang up about 0.7 miles southwest of old Robinsonville. It gained a post office in 1902 and incorporated in 1912. Greenhorn boomed between 1910 and 1915, boasting stores, three saloons, two hotels, a meat market, and a jail. The town and surrounding area, according to the 1910 federal census, counted fifty-nine inhabitants; the little city even had a waterworks system.[18]

The Bonanza and Red Boy mines began development early in the 1890s and became strong producers over the next fifteen years. At first, these two mines and associated prospects were considered stand-alone districts but eventually became part of the larger Greenhorn District. Over time, the widespread promotion of the Bonanza and Red Boy in the press helped to establish the eastern Oregon gold belt as worthy of investment. From the vantage point of 1940, the authors of the *Oregon Metal Mines Handbook* declared that the Red Boy mine was "one of the best-known mines of Eastern Oregon."[19] The mine, though discovered in the early 1880s, did not become productive until A. G. Tabor and E. J. Godfrey took it over in 1893 and installed a Crawford mill. Between 1898 and 1900, the owners put in more modern equipment such as a twenty-stamp mill, a cyanide plant, and a hoist and air compressor at a cost of over $225,000. Over time, Red Boy miners carried out more than 10,000 feet of development work on five distinct veins. By 1912, the best ore had been recovered and large amounts of water at the lower levels made further mining unprofitable. In its heyday, the Red Boy was touted by the Baker City Chamber of Commerce as "one of the most complete plants to be found in the Northwest. Operated by waterpower, lighted by electricity, heated in winter by steam, it stands as the queen of the Baker City gold fields for simplicity and durability."[20] The Red Boy Mine was also notable for introducing hydroelectric power generation to the region.

As returns from the Red Boy began to decline in the early 1900s, new owners sought more efficient ways of operating the mine. After a feasibility study, they decided in 1905 to replace the steam power system, which was run by wood-fired boilers, with lower-cost electrical energy generated by hydropower from a local water source. Following an engineer's recommendations, the mine owners formed the separate Fremont Power Company to build a

Table 5.1. Principal Lode Mines of Rock Creek, Cracker Creek, and Greenhorn Mining Districts *

Mines	Patent	Peak Production Years	Estimated Production	Year Located
ROCK CREEK				
Baisley-Elkhorn	Yes	1890–1907	$936,718	1887
Highland-Maxwell	Yes	1900–1921, 1936–1938	$625,000	1890
CRACKER CREEK				
North Pole	Yes	1895–1908	$2,485,000	1870s, 1887
Eureka & Excelsior	Yes	1894–1905	$1,064,833	1886
Taber Fraction	Yes	1903–1905	$475,200	1887
Columbia	Yes	1897–1916	$3,639,000	1888
Golconda	Yes	1897–1914	$500,000	1887
GREENHORN				
Bonanza	Yes	1892–1907	$1,750,000	1877
Pyx	No	1900, 1907–1911	$300,000	—
Ben Harrison	Yes	1913–1937	$425,000	1896, 1899
Red Boy	Yes	1890–1914	$1,000,000	1886, 1895

*Historical dollar values

The Bonanza Mine near Granite, Oregon, showing mill and accessory structures. Courtesy of the Baker County Library.

hydroelectric power plant, at the cost of $100,000, using Olive Lake as a storage reservoir. An eight-mile-long wood and steel pipeline conveyed water from the lake to the powerhouse. The Fremont powerhouse sat one mile from the Red Boy Mine and began running in 1908. It supplied power to other mines in the area in addition to the Red Boy. In 1911, the plant was sold to the Eastern Oregon Power and Light Company, which later passed the operation to another private power company. The Fremont powerhouse continued functioning until October 1967, when it was decommissioned and given to the US Forest Service. The Forest Service now maintains and manages the powerhouse, which contains all its original equipment, as a historic site.[21]

The Bonanza Mine, four miles east of Greenhorn, had a history like that of the Red Boy. Although discovered in the late 1870s, early mining efforts produced limited results until the Geiser family of Baker City took control in 1891 and then patented the mine in 1897. After extracting several hundred thousand dollars, they sold the mine in 1898 to Pennsylvania investors for a reported $500,000. In 1902, the owners installed a hoist at a cost of $125,000. During a ten-month period between January 1902 and July 1903, available records indicated the mine produced gold worth $166,627. At the same time, operators spent $21,554 on tunnel expansion. Profitable operations continued until 1905. Over time, underground workings totaled 18,000 feet of tunnels, adits, and shafts with a depth of 1,250 feet. The vein averaged 5 feet to 6 feet in width. Equipment included a forty-stamp mill and concentrator and an air compressor to operate a ten-drill plant. Laborers moved the ore from the mine shaft to the stamp mill by means of a 2,000-foot-long aerial tram. Over fifty men worked at the mine.[22]

Tables 5.2 and 5.3, reporting production values for Oregon and Baker and Grant counties, indicate that during the years 1900 to 1907, precious metal output experienced a gradual decline after peaking in 1902. Although government records for Baker and Grant counties are missing for 1900 and 1901, those for the entire state indicate that eastern Oregon output reached at least the $1 million mark, with production from 1902 through 1905, representing strong returns as well. Indeed, a writer for the *Mining and Scientific Press* declared in January 1902, "Mining in Oregon has made a substantial permanent advance during the year 1901. The districts already well established, with Baker City, Sumpter and Granite as centers, increased their gold yield and proved such an extent of new ground that a continuation of the increase of output can be looked for during the current year."[23] In a promotional piece written for the Oregon Railroad and Navigation Company in 1902, P. Donan wrote, "Of all the much-advertised 'get-rich-quick' bonanza lands of this country, none offers greater and richer opportunities to enterprising prospectors and miners than the Eastern Oregon Gold-Field . . . So far as exploration has gone, its peaks and foothills in all directions seem ribbed with ore, and its gulches and creek-beds deposits of golden sands."[24]

BOOM YEARS

At the close of 1902, the mining interests of eastern Oregon optimistically continued touting their opportunities for success. Certainly, the mining press and stock promoters looked to the future confidently. A writer for the *Pacific Miner* captured the prevailing sentiment when he stated in January 1903, "The whole field of Eastern Oregon in 1903 will be alive with brilliant mining operations, and the result in the production of gold on this continent will soon be apparent."[25] Indeed, at the end of 1902, the eastern Oregon mines did seem poised for greater success. The *Pacific Miner* reported that fifty-four mines had mills with 758 stamps in operation, with sixteen of those mines each possessing twenty or more of the ore processing machines and another twenty-five mines having at least ten each. Another factor contributing to the general euphoria of the period stemmed from a sudden burst of outside investment in the mines. Again, as a writer in the *Pacific Miner* explained:

> Last year [1902] mining in the state took on a different tone. A new feature was established—publicity. . . . A number of men who had made money in the mines of the country began to organize

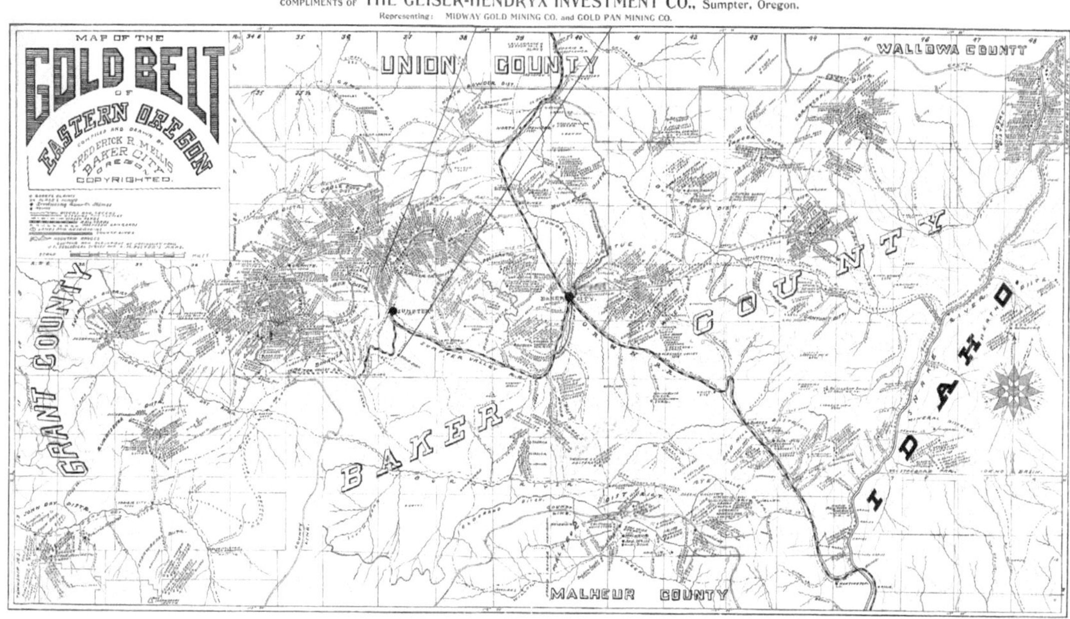

A 1904 promotional map of the eastern Oregon gold region at mining's peak of development. Courtesy of the OHS Research Library.

> and advertise. A few capitalists of New York and the East sent their agents or personally visited and investigated the country and in a few months hundreds of thousands of dollars were invested in a half dozen of the districts, with the results which electrified those who watched progress and put their capital into gold mines. . . . More fortunately still, capital met with achievement so satisfactory, so gratifying, that the groundwork was laid for enormous future development with practically unlimited capital.[26]

Indeed, official mining records for 1902 revealed that thirty-four placer and twenty-five lode mines in Baker and Grant counties produced gold and silver valued at slightly over $1.3 million (see Table 5.3).

The *Oregonian*, in its January 1904 annual edition, agreed eastern Oregon mines had great reason for optimism: "In a general way, the Eastern Oregon District is closing its greatest year, especially with reference to the many thousand feet of development done and the young properties entering the list for production. It never before saw so much good, sound scientific work as in the year 1903." The newspaper specially praised the mines around Sumpter, noting "there have been 23 new plants or material additions to old mills, giving an added capacity to the entire district of above 180 stamps." The writer pointed to the application of the latest methods of mechanical and chemical processes being used in the Sumpter mines as "representing [the] scientific application of modern improvements to a very favorable proposition in mining." The article also described the efforts to economize costs by using waterpower directly to operate machinery or employ it indirectly to generate electricity for energy—in either case, to replace expensive wood-burning steam boilers.[27]

The years 1903 through 1905 saw continued heavy investment in milling and processing equipment. In 1903, as the *Oregonian* noted in its January 1, 1904 issue, "they represented nearly every known type of some of the most modern methods of ore reduction by mechanical and chemical devices." A few of the best and longest-producing mines, such as the Bonanza, Golconda, and Maxwell, also sought greater depth, penetrating the 1200-foot level, and all sought to increase efficiency in operations. Pushing out horizontally, other mines feverously explored their underground ore veins on several levels, hoping to find the latest rich strike. These included, for example, the California, Imperial, Oregon Chief, and Last Chance in the Cable Cove District and the Belle of Baker (Mammoth), Climax, Cracker-Oregon, Ibex, and Mountain View in the Cracker Creek District. Other old mines receiving renewed attention included the Monumental, La Belleview, Magnolia, Buffalo, Cougar, and Independence in the Granite District. One of the oldest lode mines in eastern Oregon, the Virtue—near Baker City—also gained new interest. In January 1904, the *Sumpter Miner* claimed that the Sumpter Valley Railway carried between forty and fifty tons of concentrates a day from the Sumpter and tributary mining districts. A University of Oregon publication discussing the mineral resources of the state observed that mining in "eastern Oregon was quite active during 1903, and the close of the year marks a substantial advance in the general condition of things." The article went on to describe in considerable detail the nature of those advances in each mining district.[28]

A problem for investors and mine operators in the mid-decade involved the accuracy of the information about eastern Oregon mines. As a writer for the *Oregon Journal* put it in August 1904, "Oregon's mineral resources occupy the anomalous position of being unknown, yet they are well proven."[29] After a tour of the Blue Mountain gold fields, a reporter for a Chicago newspaper opined that the lack of solid data "is to be accounted for . . . by the fact that the majority of the great mines in Oregon are owned by close corporations, which go to extremes to prevent reliable information reaching the public regarding the properties they control." He then lamented the effect of rumor when honest facts were lacking: "About the best mining properties in Oregon it is almost impossible to learn exact truths, while about those which are being exploited it is difficult to keep from learning too much. Strike an average between the two . . . and the truth will not be far distant."[30]

The Chicago reporter also addressed another matter that greatly rankled the mining industry of Oregon: "The gold output credited to Oregon by the United States Government is but $1,800,000, whereas the truth is that the State probably produces no less than $6,000,000 annually, and possibly more. The discrepancy between the Government figures and the facts is partially accounted for by the efforts of close corporations to prevent a general knowledge of the facts and by the fact that a very large proportion of the concentrates is treated in smelters of foreign States [i.e. Washington, California, and Colorado], and is credited to such States."[31] J. Frank Watson, vice-president of the American Mining Congress, agreed with the reporter, stating that "it is a well-known fact that Oregon is not credited with anything like her mineral output, for the reason that the ores from the mines in this state go to smelters located in other states."[32] To the argument that Oregon did not get proper credit for its gold and silver output, the Director of the Mint had a sharp response:

> I have heard the statement made that Oregon does not get credit for mineral productions, but as a matter of fact we have complaints of that kind from almost every gold producing state in the Union. We base our figures on the returns from smelters, from the mints and assay offices, from manufacturing jewelers and from every other source which might enlighten us on the amount of gold produced. If our statements are not correct, what becomes of the bullion? Certainly no one is burying and hiding it. People do not bury it in the ground. I'm inclined to think the claims made by Oregon and every other state which produces gold are prompted by that very human desire to boom their respective districts. You can't always rely upon the statements made by the people of the mining sections as to the quantity of the output. We get all the figures obtainable from official sources and compile them.[33]

Whatever the actual amount of gold Oregon was producing, the *Oregonian* noticed the economic effect that mining had on both the state and Portland at the beginning of 1905. It observed that statewide, 3,700 men earned an average wage of $3 a day to work the mines. Including salaries of superintendents, foremen, and assayers, the mines generated a payroll worth approximately $5 million annually. The newspaper went on to argue that "the greater amount of this money is spent with Portland merchants, and the benefit thus accruing from this industry cannot be estimated." The paper further noted that "four years ago Portland had only one mining machinery house; now there are eight, besides three firms that manufacture supplies for the

Table 5.2. Oregon Gold and Silver Production, 1900–1910*

Year	Gold	Silver	Total
1900	$1,727,894	$81,866	$1,809,760
1901	$1,834,808	$98,324	$1,933,132
1902	$1,837,392	$58,015	$1,895,407
1903	$1,352,907	$67,823	$1,420,730
1904	$1,412,186	$75,284	$1,487,470
1905	$1,405,235	$54,744	$1,459,979
1906	$1,366,900	$53,162	$1,420,062
1907	$1,129,900	$57,234	$1,186,495
1908	$865,076	$23,109	$888,185
1909	$781,964	$14,470	$796,434
1910	$679,488	$19,428	$698,916

**Historical dollar values. Source: Brooks and Ramp*

Table 5.3. Gold and Silver Production: Baker and Grant Counties, 1900–1910*

Year	Baker County		Grant County		Total
	Gold	Silver	Gold	Silver	
1900	—	—	—	—	—
1901	—	—	—	—	—
1902	$1,195,417	$18,721	$74,441	$40,076	$1,328,655
1903	$702,737	$14,620	$ 102,313	$34,439	$854,109
1904	$790,828	$44,355	$82,280	$29,417	$946,880
1905	$771,607	$33,317	$88,246	$14,952	$908,122
1906	$695,653	$35,984	$55,092	$2,492	$789,221
1907	$631,045	$45,929	$80,645	$7,106	$764,725
1908	$507,929	$9,273	$88,636	$9,533	$615,371
1909	$417,539	$10,147	$41,327	$881	$469,894
1910	$401,002	$16,111	$36,201	$1,254	$454,568

**Historical dollar values. Source: Brooks and Ramp*

miner in the States of Washington, Oregon and Idaho. The shipments of mining supplies during the last year have been an important factor in Portland's trade."[34] While the *Oregonian* might have overstated the value of Oregon's gold and silver industry, the state's mines—according to the US Geological Survey and the US Mint—made a respectable showing, ranking tenth among precious metal states in 1905. That year, Oregon's 167 placer and sixty-six lode mines yielded almost $1.5 million. The average value of Oregon's lode product came to $8.03 per ton while the value for all producing states averaged only $4.82. In the eastern Oregon gold belt, Baker and Grant counties had thirty-two placer and twenty-nine lode mines reporting product valued at slightly more than $900,000; however, the two counties also had 329 nonproducing mines.[35] The potential for riches remained ever present, whether in new or old prospects or existing mining properties. The Greenhorn District seemed an especially attractive option in 1905.

While Sumpter was "better known to outside investors than any other mining camp in the state," by mid-decade, they increasingly turned their attention to the Greenhorn District, situated approximately halfway between Granite and Susanville.[36] Hoping to gain more notice for the opportunities at Greenhorn, a correspondent to the *Pacific Miner* wrote, "It is astonishing how little is heard of Greenhorn when one realizes the extent of the mining being done there." At the time he wrote, July 1904, at least six or seven mines had mills with sixty-eight stamps in operation or under construction. Both gold and silver bonanzas seemed in the offing, as mine operators and stockbrokers touted the New York, IXL, Psyche, and Morning mines as good investments and alluded to active exploratory work taking place at numerous other promising prospects. The IXL mine claimed to be yielding $30,000 to $40,000 per month in the summer of 1904, and the operators of the New York mine asserted that they had blocked out ore worth $100,000.[37]

The mining communities of Granite and Susanville also expected to benefit from the mining boom of the early twentieth century. Certainly, citizens of Granite felt emboldened enough about their future to incorporate the town, plat streets for a real estate promotion, and install a water system. At the same time, Susanville's supporters seemed sure that their mines would prove "that Susanville is a mining country not to be sneezed at. . . . While Susanville is not a city, or one of those boom mining camps which grow up in one day and perish almost as quickly, it is growing very rapidly and is as lively as could be expected." Both communities served as service centers for their surrounding mines and, in Susanville's case, small-scale agricultural operations.[38]

ARRIVAL OF THE SUMPTER SMELTER AND THE RAILROAD

The completion of the big smelter at Sumpter in 1903 and the slow but steady advance into the interior by the Sumpter Valley Railway greatly aided mining development by reducing processing and transportation costs. Mine operators also began to use waterpower to generate electricity for operating equipment at their mines. In addition, the introduction of floating dredges on Granite Creek and the North Fork of the John Day River in the early years of the twentieth century rejuvenated placer mining in eastern Oregon.

In the heady days of 1902, mine owners and other businessmen encouraged the development of a smelter at Sumpter to treat the ore from surrounding mines. As the mines increased their output, local and eastern entrepreneurs thought that a profitable smelter could be built and operated. Mine owners encouraged the idea because it had the potential to increase their profits. A local smelter would save on transportation costs, currently a big expense, since the nearest smelters receiving Oregon gold were located at Tacoma, Washington, Salt Lake City, and Denver. An observer in the *Mining Investor* speculated that the Sumpter Smelter "will be able to save the mines $6.00 per ton on their freight and smelting charges, which savings will allow many properties to produce which are now idle."[39] Another commentator in the eastern press agreed, stating that "a great quantity of ore which has heretofore been let go to waste because it was not of sufficiently high grade to warrant it being shipped some hundreds of miles to a smelter can now be treated with profit, while ore which assays at high figures will yield a better return."[40]

The success of the gamble on such a significant investment in a smelter depended on the mines consistently producing the necessary volume of ore needed to make the smelting operation profitable. A correspondent to the *Pacific Miner* assured his readers that the extensive plant under construction represented the investment and knowledge of capable "smelter men . . ." who "would [not] enter upon this elaborate scope of construction and preparatory work without careful investigations of the ore resources." He went on to affirm that "they are practical in business, practical in their profession, and they have not departed from the practical in casting their lot in the Sumpter district, where countless signs assure tremendous growth in mining."[41]

The plant, financed largely by eastern investors, covered 200 acres and stood impressively on a hillside one mile south of the town of Sumpter. Workers constructed four terraces to accommodate the various operations of the plant by removing 13,000 cubic yards of earth and rock. The smelter had an initial capacity of 150 tons of ore daily, with the ability to scale up to 300 tons. It employed the copper matting process, in which copper and iron sulfides collected and saved the gold and silver values. A large water jacketed furnace heated air used to blast and burn off sulfur and other unwanted chemicals. The hot blast furnace heated the air to a temperature of 600 to 800 degrees Fahrenheit. The smelting process utilized copious quantities of water, which were stored in two tanks, each with a capacity of 44,000 gallons. The work site had large storage bins capable of holding 1,800 tons of ore and 200 tons of coke. The main smelter building was 88 feet long, 46 feet wide, and 42 feet tall; while the powerhouse, containing engines and boilers, measured 77 feet long, 46 feet wide, and 35 feet tall. A large office building adjoining the main plant extended 52 feet in length, 32 feet in width and 12 feet in height, which also housed a drafting room and assay and chemical departments. Other structures included a sample mill—33 feet by 37 feet and 40 feet high—and various equipment sheds. An impressive smokestack standing 125 feet tall towered over the site.[42]

Construction of the smelter complex cost $225,000 and required 600,000 board feet of lumber and 500,000 bricks, which were fired on the site. Wood in the vicinity supplied the fuel to generate the electric power used in operating the plant. The Sumpter Valley Railway built a spur to connect the smelter

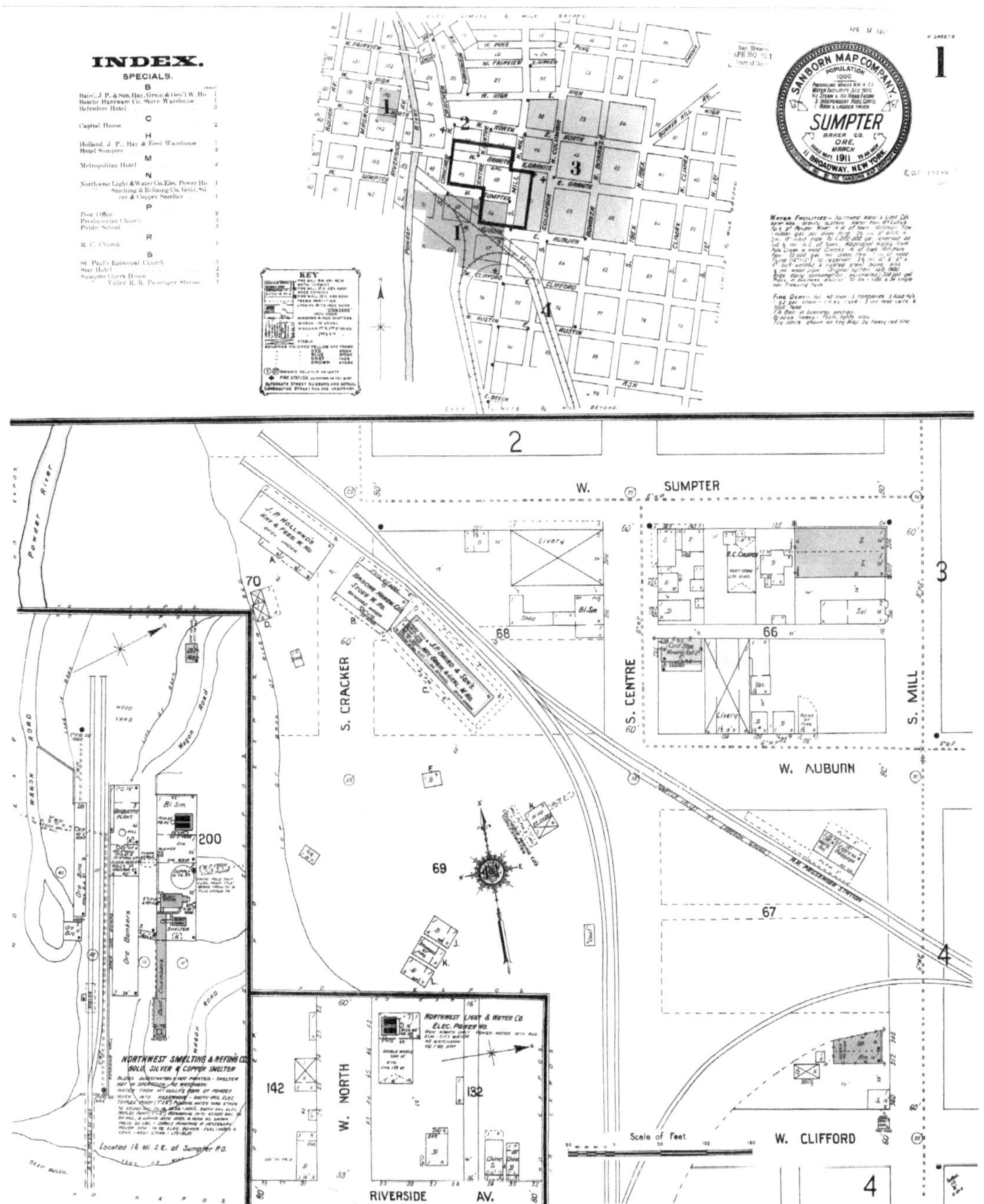

Sanborn Insurance Map, 1911, showing the floor plan of the Sumpter Smelter plant. Courtesy of the OHS Research Library.

A view of the Sumpter Smelter plant and surroundings ca. 1904. Courtesy of the Baker County Library.

complex with its main line at Sumpter. Even before the smelter went into operation, it put money into the local economy. At times, more than 100 men worked on the project, some earning $5 a day—an advanced wage for the time. Although workers had completed the smelter by July 1903, full-scale operations did not get underway for almost another year. The Sumpter Valley Railway was slow to complete the spur connection with its main line, and contracts from mine owners to supply sufficient ore to start the furnace also took time to arrange. Finally, in August 1904, the smelter was "blown in" with a force of thirty men at the start. Mine owners and managers enthusiastically welcomed the advent of the smelter. H. T. Hendryx, a Sumpter mine investor, predicted great results from the smelter's operations: "It is going to revolutionize mining in this whole camp. All roads leading to Sumpter are filled with four-horse teams hauling in ore. There is an established market for ore of all grades and the smelter has the capital and is making spot cash purchases. It is easy to see the great benefits of such an institution to a mining community." [43] The smelter ran fairly steadily until the slump in mining in 1908. According to one account, the smelter processed 19,068 tons of ore and concentrates between 1904 and 1908. Thereafter, it operated sporadically until permanently closing in the late 1920s. Knowledgeable miners questioned whether there was ever enough ore to profitably sustain such a large operation.[44]

The Sumpter Valley Railway, so important to the development of mining in Baker and Grant counties, originated as a component of the effort to exploit the vast timber resources of eastern Oregon. At the turn of the twentieth century, government experts estimated that Oregon had roughly 300 billion board feet of standing timber—one-sixth of the nation's total. While much of this timber grew in the Coast and Cascade ranges, almost 18 billion board feet stood in eastern Oregon. The main species of trees in the Blue Mountains of eastern Oregon consisted of Ponderosa or Western Yellow Pine. Once milled into lumber at Baker City, Austin, Whitney, and other nearby communities, it was sold to markets in Chicago and additional midwestern locales.[45]

In the early 1890s, the Oregon Lumber Company, started by David Eccles, John Stoddard, and others from Utah, bought large tracts of timber land in the Blue Mountains, built a sawmill in Baker City, and started construction on a narrow-gauge railroad—the Sumpter Valley Railway—to supply logs to its mill. As the company pushed deeper into the forests west of Baker City, it constructed more sawmills and eventually extended the Sumpter Valley Railway as far as Prairie City, eighty miles west of Baker City. Reaching Sumpter in 1896, the railroad not only hauled logs, but also carried passengers, general freight, livestock, and ore. From 1895 to 1910, about a quarter of the railroad's revenue came from passengers, while logs, ore, and general freight made up the rest. In the first few years of the twentieth century, the railroad annually hauled approximately 15,000 tons of ore and concentrates. In 1909, as mining decreased, the state railroad commission reported that the Sumpter Valley hauled 1,559 tons of ore concentrates.

The line was something of an engineering marvel, as the mountainous terrain it traversed required numerous horseshoe curves and spectacular switchbacks necessary to maintain roadbed grades of no greater than five percent. Building west from Sumpter, the railroad passed over the Larch Summit (el. 5,094 feet), reaching the site called Whitney in 1901, which soon had a sawmill and small settlement. Then, it moved on to Tipton Station in 1904, next to White Pine in 1904, and finally stopped at Austin in 1905. At Austin, Eccles built another sawmill and sent a feeder line down the Middle Fork of the John Day River towards Susanville to tap the vast pine forests and mining traffic of that area. Soon, a bidding war broke out between Prairie City and Susanville to see which community would get the mainline extension from Austin. Susanville offered timber and mining traffic, but Prairie City, in the main John Day River Valley, had a greater population and economic diversity, including agricultural and livestock resources available for transport. The railroad opted to build the line over Dixie Summit (el. 5,277 feet), switching back and forth down the mountain and into Prairie City, arriving in July 1910. Eccles also had a desire to extend his line from Prairie City southwest to

Lakeview, Oregon, and on to California. His death in 1912, however, ended that plan, as his brother, who took over management of the railroad, had no interest in extending to California.

The cost of transporting ore and concentrates to a smelter proved one of the biggest impediments to profitable mining in eastern Oregon. A September 6, 1901, article in the *Blue Mountain Eagle* lamented that Grant County "facilities for transportation are so poor that mining machinery and fuel cannot be taken in and the ores while of high values are not rich enough to bear the expense of long wagon hauls in addition to railroad rates." As a writer for the *Oregon Journal* concluded in 1904, "Transportation stands supreme as a factor of lode mining progress in the past, present and future. . . and today Oregon mining struggles with this problem." He went on to explain that "mountain roads, always bad, have not been assailed by mine operators and their commercial allies in a really serious mood until within the past year or two. Every prominent district is now engaged in resurveying such routes, reducing grades, improving the roadbeds and facilitating mining in a material degree." Still, as he noted, "the west's general average of 25 cents a ton a mile for mountain roads has been exceeded more frequently in Oregon hauling than reduced." He thought the ultimate solution lay in pushing further development of railroad connections to the region.[46]

THE STORY OF THE BADGER

The Badger Mine at Susanville, consisting of seven patented claims, provided a good example of the problems directly related to transportation that many mines in eastern Oregon dealt with in the early twentieth century. The previous chapter described how the initial owners of the Badger lacked the financial resources to fully develop the mine prior to 1900. After passing through one set of San Francisco investors, the mine finally came into the possession of Fred Bradley and his associates from San Francisco. Bradley, considered one of the leading mining engineers of the time, also owned silver mines (Bunker Hill and Sullivan Company) in northern Idaho and had a part ownership of the Tacoma smelter. He paid $500,000 for the Badger Mine.[47]

When Bradley bought the Badger mine in 1902 and patented it, the operation consisted of a ten-stamp mill, a cook and boarding house, an office, bunkhouses, ore storage buildings, and a hosting plant with a forty-horsepower engine. This collection of buildings and equipment represented a typical setup for a lode mine of average production. Between 1899 and 1901, net profits came to about $109,000, and the mine usually employed between thirty and fifty workers, including a few women as ore sorters. Upon taking possession, Bradley pushed development work underground, especially at the 500-foot level below the surface. All this effort yielded large amounts of

Above: the Badger Mine at Susanville, Oregon, ca. 1905. Courtesy of the Grant County Museum.

Below: 1900 General Land Office map of the Badger patented mining claims. Courtesy of the Bureau of Land Management.

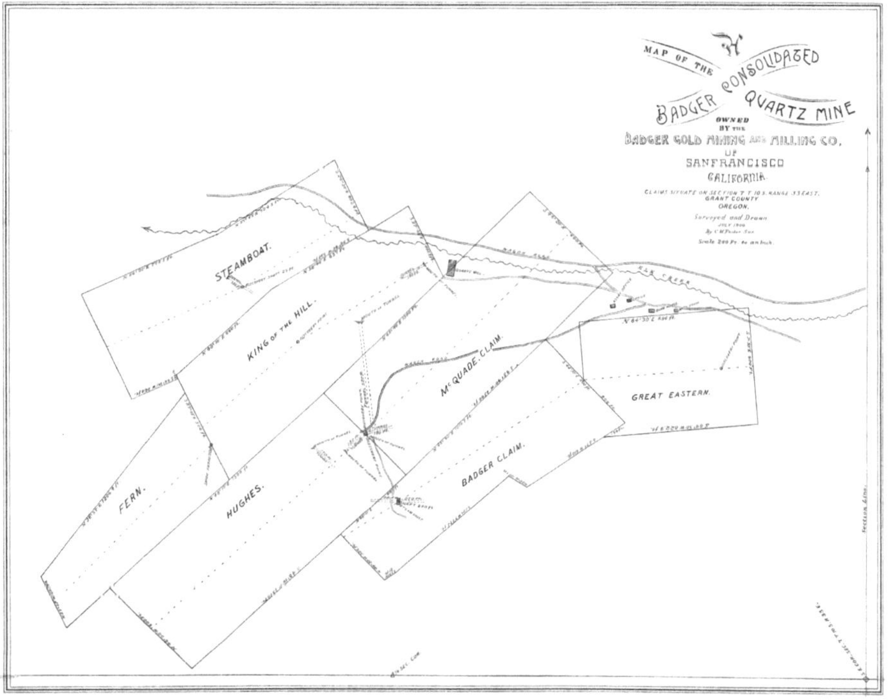

high- and low-grade ore. He next added an aerial tramway from the mine to the mill and a concentrating plant to process the low-grade ore and installed a new mine superintendent, E. P. Kennedy.

Once milled and chemically processed, transportation of the ores and concentrates to a smelter became a cumbersome and expensive undertaking. As one observer noted, "The ores are concentrated and shipped, the matter of hauling the concentrates from this mine being quite an industry in itself."[48] The Pendleton *East Oregonian* reported that one such trip between Susanville and Pendleton took "two large and strongly built wagons" to haul 9,420 pounds of ore "in small sacks, each about the size of a 50-pound flour sack. . . . Each wagon hauled 40 sacks, so that the estimated value of the two loads was in round numbers $6,000." The reporter went on to note that the "ore is stored here in a warehouse until a carload has accumulated, when it is shipped to the smelter." He also remarked that "several more trips will be made by the transfer men this summer, and they carry back with them provisions and supplies of various kinds. . . [as] the mines draw a great deal of their supplies from here [Pendleton]."[49]

Before the completion of the Sumpter smelter, the Badger's rich ore and concentrates had to be hauled by horse-drawn wagons eighty-four miles north, over tortuous mountain roads to the railhead at Pendleton for shipment on to the Tacoma smelter via the Oregon Railroad and Navigation line (Union Pacific). The alternative route required hauling the mine production sixty miles east over equally mountainous terrain and poor roads to Sumpter. There, the Sumpter Valley Railway would transport the ores on to Baker City and then transfer the shipment to the Oregon Railroad and Navigation line for carrying to Tacoma. As the Sumpter Valley built its line further west after 1901, the transportation costs were reduced but remained a major expense. For example, during 1903, it cost $13.95 per ton to haul to Whitney (the nearest station on the Sumpter Valley Railway) and $17.57 per ton to transport to Pendleton. As a mine report at the time noted, "Had the railroad been extended to Susanville there should have been a saving on hauling of at least $12 per ton." Since Bradley owned an interest in the Tacoma smelter, he stood to profit from both the mining and smelting of Badger ores.

In 1904, however, the Badger mine began sending its product to the new smelter at Sumpter. By then, the Sumpter Valley Railway had reached Tipton—only thirty miles from Susanville. At this time, the Badger employed eight, six-horse teams to transport the ore and concentrates from the mine to Tipton. Unfortunately, just as transportation costs came down, the mine itself became less productive. Between March 1902 and December 1903, the mine received $107,000 from the smelter but declared expenses of about $140,000; the ore cost $33,000 more than it returned. Much of the loss resulted from the high transportation costs. In 1904, with lower transportation costs, the mine returned a profit; the company even installed a new aerial tram system and a water-powered air compressor plant to service both the Badger and the Bull of the Woods—a companion mine. In 1905, however, the operation again lost money and closed in November 1905. Discovery efforts below the 500-foot level had disclosed the presence of paying ore in sight, but at that point, legal troubles with an adjoining mine convinced Bradley to shut the Badger. The owners of the neighboring claim, held by the Stockton Gold Mining Company, sued the Badger Gold Mining Company, asserting that it had entered the Stockton claim underground without permission and removed valuable ore. The court found in favor of the Stockton company's legal position but left the matter of damages to a future suit. The litigation was reported to have been settled in 1914, but mining did not resume until 1927 and then only briefly.

CHALLENGES TO SUCCESSFUL MINING

The Badger story also points out how litigation could upend even a successful mine. Some of the biggest and richest mines, such as the Golconda, Red Boy, Bonanza, La Belleview, and the Eureka and Excelsior all experienced extended closures over lawsuits brought by competitor mines, disgruntled stockholders, or unpaid suppliers and employees. The issues included assertions of illegal encroachment or faulty titles, nonpayment of dividends, attachments for unpaid bills, and the breach of sales or lease agreements.[50]

Another impediment to sustaining a successful mining operation involved the difficulties inherent in raising funds for mine development. Opening a mine could be an expensive proposition. One experienced miner estimated it could cost anywhere from $50,000 to $1,000,000 to develop a paying mine.[51] Another knowledgeable mine operator told the *Pacific Miner*:

> Beyond question, one of the main elements of success in ledge mining is capital and plenty of it. While farming, manufacturing, merchandising and many other branches of industry may begin in a small way with a limited capital of but a few hundred dollars, and gradually develop into great enterprises, ledge mining nearly always requires the expenditure of many thousands before a point is reached where the business becomes even self-supporting.

He went on to state: "In most lines of industry it is possible to realize small profits from the start, but in mining, especially in low grade, rebellious ores, profit comes only after the expenditure of what may be considered a fortune." He added, "Hence the financiering of a mining proposition becomes one of the chief features in the success of the enterprise." He concluded that unfortunately "the history of all mining is that local capital, either through timidity or indifference that usually obtains regarding opportunities close at hand, rarely, if ever, takes the initiative in mining enterprises."[52]

Raising the necessary funds for such a speculative undertaking often opened the door to unscrupulous stock promoters. An editorial in the *Portland Telegram* put forth a theory that three types of mining promoters operated, like the three types of ore: high-grade, low-grade, and base. According to the editorial, the "the high-grade promoter is the man who has an eye single to the interest of his purchasing client; who is careful to deal in nothing but first class properties, and in the exploitation necessary to their disposal or further development, adheres strictly to facts." On the other hand, "the low-grade promoter is disposed to indulge in what may be considered as a somewhat legitimate game of chance. The financial ventures that are offered to the public by this enterprising genius are based upon conditions that may result either in profit or loss. The first requisite with this class of promoters is . . . a lively imagination, with a conscience that will not interfere in the preparation of a prospectus promising to the 'lucky investor' the earth and the fullness thereof." Finally, the editorial stated, "the base promoter is to be dismissed with a sentence, and that a brief one. He is a swindler, pure and simple." The editorial further lamented, "It is safe to say that Oregon has not suffered in this regard [dishonest promotion] so extensively as some of her sister States; but even here the evil of conscienceless promotion has been altogether too apparent. There is no doubt that the reputation of Oregon mining has suffered on this account." The result had been, according to the editorial, that the "meritorious project finds difficulty in making legitimate headway, and the general industry is not accorded the public consideration to which it is entitled."[53]

A correspondent to the *Oregon Journal*, in a survey of eastern Oregon mining in the summer 1907, also lamented the baleful effects of dodgy mine promoters. He decried "unscrupulous wildcatters posed as mine managers, and where one dollar was spent in the ground too often two went into the promoter's pocket and the work which was done was simply to make a showing without regard for what might result in the way of taking out the rich ore which underlies the whole of the eastern Oregon . . . gold belts." This writer, however, optimistically wrote that the "mad season" of the wild cat boom era in eastern Oregon had given way to the solid investment and operations of serious capitalists.[54] As late as 1912, though, the director of the State Bureau of Mines, H. M. Parks, lamented that "a few years ago mining received a setback, due largely to the promotion of questionable projects that led capital to be unusually cautious about investment in mineral properties."[55]

Bourne, Oregon, ca. 1910, showing the town's relationship with the Cracker Creek mines. The Columbia Mine is in the center, and the E&E Mine is in the upper portion of the draw. Courtesy of the Baker County Library.

Unfortunately, the mature eastern Oregon mining scene was no stranger to the "base" mining promoter. The mining stock promotions of Letson Balliet represented a particularly egregious example of the worst swindlers. Balliet arrived in Baker City about 1900 and purchased a controlling interest in the bankrupt and abandoned White Swan Mine. In November 1900, he bought the weekly *Baker City Herald* and turned it into a daily. He then began advertising the mine in glowing terms in newspapers throughout the country, selling stock in the White Swan on a monthly payment plan. Balliet claimed that through his previous mining activity, he had amassed a personal fortune and made millions for his shareholders. His salesmanship eventually raised $300,000 from 1,200 small investors across the United States. When the investors failed to receive any dividends from the White Swan Mine and Balliet's extravagant lifestyle came to light, federal authorities investigated and charged him with mail fraud. After several trials, he was finally convicted but received only a small fine and a minimal jail sentence.[56]

That Balliet got off so lightly for his thievery angered the *Pacific Miner*. The editor of that journal wrote that the light sentence "does not appease the multitude of poor people who lost their mite through Balliet's criminal work and grossly fraudulent advertising. . . . Successfully baffling of the law for four years after the mesh first closed around him, and then such petty atonement as Balliet has given, will be the best argument on earth for the next criminally inclined sharp to follow his footprints."[57] An even more extravagant example of what the *Pacific Miner* feared was already at work in the mining center of Bourne.

In 1900, the Cracker Creek mining community of Bourne was booming, named in honor of the Portland politician who owned the district's famous E&E mine. The area's numerous rich mines had attracted wealthy investors from across the country and even Europe. Mining experts stated that the Cracker Creek district's well-defined gold veins would continue to great depths. But successful mining at great depths required the application of large infusions of capital and new mining technology. For a few years, Bourne's mining companies and mining stock promoters found it easy to entice investors. Purchasers eagerly bought Oregon mining stocks, and Bourne enjoyed the good times. The 1900 federal census listed almost 600 residents, although ten years earlier, the gold rush boomtown did not exist. In its heyday, the town boasted three hotels, six saloons, a newspaper, and many stores. Business interests even promoted building a railroad that would cover the six uphill miles from Sumpter to Bourne. After 1910, mining in the vicinity of Bourne slowed but continued at a low level; however, a flash flood in 1937 wiped out most of the town's buildings.[58]

While legitimate mines and prospects existed around Bourne, unscrupulous wildcat promoters also set up shop in the town. These fraudsters hawked and sold stock in nonproducing mines, scamming thousands of innocent investors. The most famous and mysterious of the stock swindlers was F. Wallace White. After arriving in Bourne about 1903 from Ohio, White started the Sampson Company Ltd., with offices in Bourne, New York, and London. He then set up a newspaper with two separate editions: One, for local consumption, contained legitimate news; another, for worldwide distribution, reported fabulous mining success stories and advertised stock shares at $1 or less in mines with little or no production records. White's Sampson Company ambitiously advertised that it would buy all the big-producing mines in the Cracker Creek district and merge them under one ownership. He may have raised as much as $5 million through his various promotions. To provide a façade of respectability and financial success, he built a hillside mansion with expansive terraced grounds overlooking the mining town. There he entertained his guests with lavish dinners and fancy balls. After about ten years, he began to exhibit signs of unease as he learned that federal authorities were investigating him for mail fraud. One night, he and his wife skipped town with his money, leaving everything else behind.[59]

White's hasty departure from Bourne coincided with the beginning of the town's decline. In 1900, the Bourne census precinct had 592 inhabitants; by 1910, it counted only 220. The mines in the district and elsewhere in the eastern Oregon gold fields had become less productive while the expense of deep mining had increased. This situation began unfolding just as the nation experienced a major financial panic. Over-speculation in many industries, including mining, and the economic effects of the San Francisco earthquake in 1906 had destabilized the national economy and set it up for a sharp downturn. A banking failure provided the catalyst for a general economic collapse in October 1907, triggering the first worldwide financial crisis of the twentieth century. The financial collapse happened quickly: Caught up in an unsuccessful attempt by speculators to corner the copper mining market, the Knickerbocker Trust Company of New York failed. This event, in turn, caused runs by depositors in other banks and led to a collapse in the New York stock market and a general tightening in bank lending. The brief but sharp national economic contraction and subsequent reduction of lending made it difficult to raise financing even for sound mining operations. The financial stress also caused the Sumpter Smelter to close. The eastern Oregon mining boom was over.[60]

The financial chicanery involved in the promotion of eastern Oregon mining stocks reflected a problem endemic to western mining in general. One wag went so far as to define a mine as "a hole in the ground with a liar on top."[61] Western historian Elliott West concluded that "the combination—greater potential wealth, more money needed, and greater risks—made mining more economically volatile than any other western enterprise. It also left it marvelously open to the baldest manipulation and the wildest chicanery. As such it was arguably the nation's clearest instance of the plunging spirit so much a part of late nineteenth century America."[62] In assessing the problems inherent in the relentless search for mining capital, historian Richard White noted that "many promoters," finding "it was far more profitable to mine investors than to mine ore . . often pocketed up to 50 percent of the capital they raised and the value of the capital stock of mining companies often far exceeded either their assets or their potential profits."[63] In fact, overspeculation affected gold and silver mining worldwide.[64]

DECLINE

While mining in eastern Oregon did not completely stop, it did significantly slow down after 1907. In 1905, Baker and Grant counties produced gold and silver worth $908,122; in 1907, output dropped to $764,725; by 1910, it stood at only $454,568—half what it had been in 1905 (see Table 5.3). The slowdown was notable, particularly at many of the larger mines. For example, in the Cracker Creek district, the North Pole's production for 1907 came to $203,557, but in 1908, it plunged to $68,258. The suspension of operations at the fabulously rich North Pole mine struck a particularly ominous note for the district. Its record of production between 1895 and 1908 had been impressive, crushing and treating 158,917 tons of ore to yield gold and silver worth $2,485,006. It typically employed seventy-five men and had an extensive plant.[65]

After 1908, all the major Cracker Creek district's mines ceased operations except for the Columbia Mine, which continued until 1916. In the neighboring Elkhorn district, the Baisley-Elkhorn mine experienced a production decline from $210,000 in 1905 to $38,481 in 1907 and then virtually zero in 1912. In the Greenhorn District, the Red Boy steadily fell from a high of $186,401 in 1900 to $10,000 in 1908. On the other hand, a few mines managed to do well. For example, at the Highland mine, production rose from $7,781 in 1905 to $111,472 in 1913. While rich, the ores from the main Cracker Creek vein required expensive amalgamation and cyanidation to process. Investors had to gamble on the prospect of continued development finding additional ore that would pay the high cost of discovery and processing. Another source of gold after 1905—again at a lower level of production—came from hydraulic and dredge mining in the region's streams and rivers. But the new mechanical forms of placer mining developed slowly and had their greatest impact after the First World War. During the slow times after 1907, small mines and prospects especially suffered from the lack of financing. Operators of unpatented mining prospects could only afford to perform the required annual assessment work necessary to legally hold their claims.[66]

Some eastern Oregon mining experts tried to put the best face on the overall decline in production from the region's mines. In a dispatch to the *Oregonian*, the president of the Sumpter Development League stated that "at the close of the year 1907 the condition of the mining camps in Eastern Oregon . . . is much more satisfactory than at any time in the past three or four years." As proof of this belief, he noted that the smelter in Sumpter "has been able to run for a period of several months longer than any time since it was established." He asserted that "ore for the smelter had come from four or five big producing mines and some 30 or 40 smaller ones. . . . The smaller properties in the aggregate have produced probably as much as the larger ones." Moreover, he looked with confidence to the future because the Fremont Power Plant, built by the Red Mine owners, had come online in October 1907 and promised to supply cheap hydroelectric power to all the mines in the region around Sumpter. As he phrased it, "The introduction of electricity will do away very largely with the extensive equipment involved in steam plants and also permit prospectors to tap the wires and get their properties into shape without the necessity of heavy expenditures for machinery." In addition, the steady expansion of the Sumpter Valley Railway westward promised to lower transportation costs. Finally, he pointed to investments in new prospects and continued development at old properties.[67]

In the Spring of 1909, the local press once again touted the mining prospects of Grant County. In a lengthy article, the *Blue Mountain Eagle* declared: "Mining is coming into a new life in Grant County. Evidence of this fact is now apparent on all sides. The revival of this great 'industry' is not coming as a boom but is progressing along the most conservative lines and is based on sound business principles." A reporter for the newspaper, after a trip through the county's various mining districts, assured his readers the days of "wild-cat and the get-rich-quick schemers" was over and that "mining as it will now be conducted will be similar to any other business." While admitting that the lack of capital and high transportation costs still hampered profitable mining in the county in general, the reporter pointed to the renewed activity at Susanville where "many miners and prospectors are at work in every direction." The Susanville miners had both placer and lode operations underway, although he confessed that the Badger Mine, the district's best producer, had not resumed operations because of litigation and high transportation costs. Still, he observed that "there is more work now going on in the quartz mines than has been for several years past." The reporter claimed that the Ophir, Gold Bug, Cabell and North Gem—each employing eight to ten workers—were yielding "returns . . . in sufficient quantities to pay for all the development work." The reporter concluded his piece with the grand assertion that once the transportation issue was resolved, "Grant County will establish herself in the commercial world as a mining district to be reckoned with and will attain a position of first rank in mining circles."[68]

Indeed, over the next few years, sporadic exploration work continued at the major mines of Baker County in the Cracker Creek district, even though the Columbia Mine was the only one in regular ore production. As the *Mining and Scientific Press* observed on July 24, 1909, "At present the Columbia is the only producing mine in the district, but the suspension of operations in most of the others is believed to be temporary and not caused by exhaustion of the orebodies." Looking back from January 1910, the same mining journal reported on conditions in Oregon during the previous year: "The first half of 1909 was exceedingly quiet in the northeastern part of the State, but several old properties resumed operations later, a few were brought into production, and new capital was interested in other cases."[69] Even the noted geologist, Waldemar Lindgren, in a survey of western mining written in February 1910, observed "a marked decrease is indicated for Oregon."[70] The *Oregonian*, however, remained ever optimistic about mining in eastern Oregon, reporting in

February 1911, "Now the wildcat speculator has passed into history and mines are being operated upon their merits. . . . The mining outlook grows brighter every year."[71] A consulting engineer writing for *The Engineering and Mining Journal* agreed with the *Oregonian*. He noted that during 1910, several mines surrounding Sumpter and Susanville had carried on extensive development and that the Sumpter Smelter had once again resumed operations. Still, he closed his report with a sober assessment:

> While the closing of 1910 did not show as extensive an increase in mining activity as might be desired, this lack was in a great measure compensated for by the solidity of such organizations and resumption of operation that took place, and which by economical and intelligent management it is contemplated will demonstrate the intrinsic value of eastern Oregon properties.[72]

Determined optimism about future mining possibilities even found its way into the state's official publication, the *Oregon Blue Book*, launched in 1911. According to that publication, "the most encouraging feature in the mining situation in Oregon consists of the fact that 'wild-catting' has been almost entirely eliminated from the industry in this State, and money that formerly went for promotion, flotation, commission, brokerage and graft, is now being legitimately applied to the opening up of mines and developing ore bodies of commercial grade." Still, in the same issue of the *Blue Book*, the state commissioner of labor warned prospective emigrants that "while Oregon's undeveloped resources are truly great and all that is claimed for them by her many enthusiasts, . . . there are no bonanzas of gold or silver merely awaiting the eye of the casual observer. All such have long ago been staked and prospected, and like many other eldorados, they are largely a thing of the past to the man with the pick and shovel."[73]

In an essay from the vantage point of 1914, a geologist with the US Geological Survey, James Pardee, provided a balanced evaluation of the mining conditions in the Sumpter area. He noted that of the lode mines, only one was in steady production while another five intermittently turned out ore. Three other mines had development work ongoing at a moderate level. Another forty or fifty prospects were carrying out their required annual assessment work. Still, he wrote that "contrasted with the activity of former years, the present relative depression is very noticeable." Pardee, however, tried to end on an optimistic note. He concluded:

> An ideal condition of the mining industry in any locality is attained when its mines are judiciously worked to their fullest capacities.

> So numerous are the obstacles to be overcome, such as uncertain ownership, lack of capital and engineering skill, incompetent management, etc., that few mining districts reach or maintain the ideal condition. Although knowledge of the Sumpter region is far from exhaustive, the facts that have been ascertained lead to the conclusion that, aside from possible new discoveries, if a few of the larger idle mines were added to the list of producers it would enjoy about the ideal condition described. . . . The continued life of the mining industry in the quadrangle [Sumpter] depends upon new discoveries and continuation in depth of the known ore shoots. . . . As an occasional new discovery of importance may be expected, the conclusion is drawn that a healthy degree of mining activity can be maintained in the Sumpter quadrangle for several years to come.[74]

Writing in 1941, Oregon mining expert Albert Burch offered yet another reason for the decline in eastern Oregon mining activity after 1908. He noted that the older major mines had been largely worked out and few new discoveries made. He observed, "It must be remembered that miners, as well as prospectors, are migratory and that they readily desert old fields for greener pastures elsewhere. There was boom on all over the state of Nevada and in Ontario Province of Canada and these booms attracted thousands of miners and hundreds of investors and promoters from all parts of the world, including Oregon."[75]

Published reports by other federal agencies confirmed the observations of knowledgeable mining engineers and geologists. In 1902 and 1909, for example, the US Census Bureau released data that surveyed the nation's mines and quarries. These studies provide a convenient statistical comparison of Oregon's precious metal mining at its peak and bottom, placing it in a national context as well. In 1902, Oregon had 262 mines, producing gold and silver valued at $1.8 million. Mining activity also generated another $5.4 million in related manufacturing activity, which was 12 percent of the total value of all manufacturers in the state. By 1909, the number of producing mines (both lode and placer) had dropped to 129, with 288 lying idle. The value of the output had declined to $627,734. The 1909 report also detailed the type and number of mine operators:

	Lode Mines	Placer Mines
Individual owner	5	31
Firm	5	17
Corporation	22	12
Total	32	60

In 1902, the mines employed 855 wage earners who averaged $955 in yearly pay. Another 110 salaried employees made, on average, $1,332 yearly. While Oregon miners accounted for only 6.3 percent of the combined manufacturing and mining wage earners, they earned 11 percent of the wages. During 1909, the Census Bureau found that the Oregon mines employed 613 wage earners whose pay averaged $656, while sixty-three salaried personnel averaged $1,305 a year. Additionally, the 1909 report noted that 99 percent of lode miners worked an eight-hour day, while only 31 percent of placer miners labored at that rate. Sixty-three percent of placer miners worked ten- or eleven-hour days. In both time periods, Oregon's wages and salaries for those engaged in gold and silver mining were comparable to those on the national level. In 1902, Oregon ranked ninth in the nation for gold and silver production; by 1909, it had dropped to tenth. In comparison with other western mining states in 1900, Oregon's miners made up a smaller percentage of the working population: Oregon, 3.3 percent; California, 4.2 percent; Montana, 15.2 percent; Nevada, 13.8 percent; Colorado, 13.7 percent; and Idaho, 11.7 percent.[76]

Between 1900 and 1910, precious metal mining played an important role in Oregon's booming economy. Over the decade, the total assessed value of the state's taxable property increased exponentially:

1900	$117,804,874
1905	$309,256,689
1910	$694,727,631

Much of that increase, however, resulted from the higher assessed value placed on railroad property and timber land after 1905. In the early twentieth century, the state derived its wealth from exploiting the bounty of its natural resources, such as minerals, timber, agriculture, and fisheries. As Table 5.4 shows, gold and silver production contributed a higher portion of the economic value of the leading resources at the beginning of the decade than at the end. Over time, timber became the leading source of wealth for the state. Still, during the decade, gold and silver generated almost $15 million for the state's economy. Occupation numbers as tracked by the Census Bureau also confirm the economic shift in resource development over the decade. In 1900, miners accounted for 3,910 workers, while by 1910, their numbers had shrunk to 2,509. On the other hand, Loggers and sawmill workers increased from 5,130 in 1900 to 14,621 by 1910.[77]

Table 5.4. Value of Oregon Products, 1900–1910 (in millions)*

	1900	1905	1910
Timber	$8.8	$18	$30
Gold & Silver	$1.8	$1.5	$699,000
Wheat	$8.6	$9.1	$11
Wool	$2.5	$3.0	$3.8
Salmon	$2.3	$1.7	$2.1

**Historical dollar values*

POPULATION CHARACTERISTICS OF GRANITE AND SUSANVILLE

The demographic profile of Susanville and Granite during the first decade of the twentieth century differed considerably from the situation between 1870 and 1890. In the earlier decades, both communities manifested placer mining operations, while in the later period, quartz mining dominated. Prior to the 1890s, small Euro American settlements shared the ground with large Chinese populations. The federal census records help to track the transition to more dominant Euro American communities as the Chinese left the area. The census data also captures many demographic features of Susanville and Granite as lode mining pushed both districts to the peak of precious metal mining and then to its bottom. While initially placer and hard rock mining dominated economic activity in both places, the Susanville precinct had also developed an agricultural component in the 1890s and early 1900s. By 1898, according to the *Oregonian*, "The farmers on the middle fork of the John Day river furnish an abundance of vegetables and fruits, and game is plentiful in the surrounding mountains."[78] The expansion of Susanville's ranching segment meant that its population experienced less turnover than occurred in Granite.[79]

Between 1888 and 1899, along twelve miles of the Middle Fork of the John Day River (in T10S, R32E and R33E), nine settlers received homestead patents of 160 acres each. Another homestead went to patent in 1902. In 1891 and 1892, two settlers made cash entries for 160 acres apiece. In the 1900 federal census for the Susanville precinct, the data recorded twenty farmers, eleven farm laborers, eight sheep herders, and two stockmen. Altogether, those involved in agriculture made up 25.9 percent of the occupations listed (41 of 158), while the mining sector accounted for 57.6 percent of the total (91). The specialized mining positions included three superintendents, two stationary engineers, one civil engineer, and one gold sorter. The remaining occupations included business (seventeen), professional (three), and miscellaneous (five) activities. The Chinese portion of the precinct included ten miners, two merchants, one cook, and one laundryman.

The ethnic makeup of Susanville in 1900 consisted of fourteen Chinese and 262 whites. The Chinese were all males, while the white population listed 182 males (70 percent) and eighty females (30 percent). For those with a place of birth noted, fifty-one (18.6 percent) were foreign born while 223 (81.4 percent) claimed the United States as their place of origin. Of the native-born inhabitants, only eighty-five (31 percent) gave Oregon as their place of birth. The census records also provide information on household and family structure. For example, the Susanville precinct had ninety-eight households with an average size of 2.8 individuals, and only two females were listed as head of household. The Euro American households were comprised of forty families, averaging 4.1 persons each. Twenty-nine nuclear and eleven extended families made up this group of households. Single men or single, unrelated men living together in various combinations—usually two to a household—made up the remainder of Euro American living arrangements.

Susanville in 1900 was a relatively youthful community of 276 inhabitants, with males having an average age of 27.14 and females, 20.1. The precinct had eighty-two children under the age of 18. The Chinese inhabitants, on the other hand, had an average age of 46.6, with the youngest being 25 and the oldest 64. The Chinese living arrangements included three living in single households, two in two different households, and seven together in another household. One-half of the Chinese inhabitants could speak English. Perhaps indicative of the *kongsi* household arrangement (described in chapter six), out of seven individuals present in the larger group, only the head of the household could speak English.

The 1900 federal census reported a new category of information: home ownership. The census form asked whether a property was a home or a farm, and whether it was owned or rented. If owned, did it have a mortgage? No property in Susanville had a mortgage on it. The schedule listed fifty-eight property owners, twenty-three having farms and thirty-five possessing homes. Six individuals rented farms, and thirty-four rented homes. Unusual shelter arrangements also existed: two miners called a wagon home, and for nine other miners, five tents served as homes (as three lived singly and two other tents held three each). In addition, another tent accommodated a family of six Euro Americans plus one boarder. The head of household and his boarder were common laborers, according to the census taker. A saloon keeper also lived in a tent while seven sheepherders lived in three other tents (as one tent had three and two held two persons each). In total, twenty-four individuals called a tent home.

The Susanville census for 1910 counted 280 inhabitants and revealed the community's continuing economic transition from mining to agriculture. While the overall population remained stable from 1900 to 1910, the number of prospectors (twelve), miners (thirty-five), mine managers (two), assayers (one), and engineers (seven) engaged in placer and lode mining declined from ninety-one to fifty-seven. At the same time, agriculturalists—including farmers (thirty-eight), farm laborers (nine), and stockmen (one)—rose slightly from forty-one to forty-eight. Mining-related occupations then made up 44.5 percent of the total, while agriculturalists comprised 37.5 percent. Twenty-three other occupations (17.9 percent) included merchants (four), teachers (four), housekeepers (three), carpenters (two), and miscellaneous professions or trades such as forest ranger, government land surveyors, blacksmith, telephone operator, railroad locomotive fireman, hotel manager, shoemaker, store clerk, and logger (ten).

The ethnic makeup of Susanville in 1910 consisted of eight Chinese and 272 Euro Americans. The total population included 169 males and 111 females, of whom twenty-nine were foreign born (10.4 percent), and 138 (49.2 percent) listed Oregon as their place of birth. The white population consisted of 161 males (59 percent) and 111 females (41 percent). The remaining 113 (40.4 percent) claimed various other states as their place of origin. The census listed 114 households with an average size of 2.45 individuals. The census taker listed only nine females as heads of household. The census also recorded fifty family units, averaging 4.2 persons each; forty-four of these were nuclear and six were extended in composition. The rest of the households consisted of either single individuals or two or more unrelated persons residing together. The statistics for family size and composition in 1910 showed little deviation from those reported in 1900.

Susanville citizens, by 1910, had a slightly older age composition than that of ten years earlier. Males averaged 35.52 years old in 1910 versus 27.14 in 1900, while females averaged 23.70 years in 1910 as opposed to 20.1 previously. The community had eighty-seven children younger than 18, about the same number (eighty) as in 1900. The eight Chinese in Susanville had an average age of 52.47, while in 1900, the average for the Chinese stood at 46.6. The oldest Chinese male was 74, and the youngest, 12. Three of the Chinese residents worked as placer miners, two others operated a grocery store, and an additional male who had a wife and young son ran a general store. The husband had immigrated in 1882, and the wife in 1881, and their son had been born in the state of Washington in 1898. The eight Chinese in Susanville represented 22 percent of Grant County's Chinese inhabitants in 1910.

The 1910 census continued asking information about property ownership. The census taker recorded that the precinct had forty farms. This represented an increase over the twenty-three noted in 1900, indicating the continuing transition to an agricultural economy. While no farm had a mortgage on it in 1900, five did carry one by 1910. The number of homes had almost doubled as well over the ten-year period: from 35 to 62. Renters occupied twenty homes; only one farm was a rental.

When the census taker arrived at Granite in June 1900, he encountered a boom town and thriving mining district. By that time, placer mining had produced approximately $2 million in gold, and extensive lode mining had recently gotten underway. The general revival of the eastern Oregon mining industry in 1897 led to the location of twenty-five quartz claims in the Granite District by 1900, with eleven lode mines going to patent by 1906. The success of the Red Boy, La Belleview, and other mines seemed to portend a bright future. In the early 1900s, the town of Granite prospered as the supply point for the surrounding mines, and the state legislature granted it incorporation in 1900. James Tabor, a long-time resident of Granite, later wrote that in 1900, the town had two large hotels, three general stores, five saloons, a drug store, a livery stable, a blacksmith shop, and a sawmill. The town also boasted two newspapers, a school, and a gravity water system. The future looked bright for Granite, as there was even talk of building a railroad between Sumpter and Granite.[80]

The 1900 census reported that the Granite Precinct had a population of 591 and, separately, the town of Granite counted another 245—making for a total of 836 in the community. The larger precinct had 189 households, averaging 3.12 persons each, while the town possessed 84 households, averaging 2.91 persons each. The Granite Precinct counted 453 males (average age, 34.25) and 138 females (average age, 22.81). In the town, males numbered 150 (average age, 29.26), and females, 95 (average age, 22.69). For the combined Granite communities, the gender composition consisted of 72 percent males and 28 percent females. The Granite Precinct had ninety-two children under the age of 18, while in the town, this segment of the population numbered seventy. The higher proportion of children under 18 in the town (29 percent) as opposed to the percentage in the surrounding district (16 percent) helps account for the younger average age of the town population.

The census taker enumerated ninety-four foreign-born (16 percent) and 497 United States-born (84 percent) residents in the Granite Precinct. Ten of the foreign-born were Chinese, twenty-four Canadian, fourteen German, ten Swedish, ten British, and seven Irish. The remainder consisted of scattered national origins. For those with the United States as recorded place of birth, Oregon accounted for only 21.78 percent of the total (108). In the town of Granite, the census schedule counted twenty-six foreign-born (10 percent) and 219 of US extraction (90 percent). Only three Chinese males resided in the town. In 1900, Susanville and Granite precincts together accounted for 24 percent of the Chinese present in Grant County.

As would be expected, mining-related occupations accounted for 67 percent of the listed occupations (278) in the Granite Precinct, while 33 percent held service or commerce types of employment (138). Of the ten Chinese residents in the precinct, seven worked as miners and three as cooks. In the town of Granite, the proportions doing mining-related work and service employment were, as might be expected, almost the reverse of the surrounding district. Only 32 percent (forty-four) worked in mining, while 68 percent (seventy-one) held service or commerce-related jobs. The three Chinese inhabitants labored as cooks (two) and laundryman (one). The town also numbered two prostitutes among its employed workers. The specific functions performed by the mining community in the combined town and district precinct reveal how sophisticated the lode mining operations had become. Besides 275 general miners, the other listed mining-related occupations included eight mine supervisors or managers, eight assayers, seven engineers of various types, four machinists, three quartz millmen, two electricians, and various miscellaneous jobs such as foreman, amalgamator, timberman, surveyor, and draftsman.

By 1910, the population of the Granite Precinct and town had become a mere shadow of its former self. The precinct dropped from 541 to only fifty-nine, while the town decreased from 245 to eighty-nine. The precinct contained forty-six males (average age, 41.28) and thirteen females (average age, 32.07). The town, on the other hand, had forty-three males (average age, 31.77) and forty-six females (average age, 23.11). The gender composition for the combined population of the communities was 60 percent male and 40 percent female. The precinct had thirty-four households, each with an average of 1.7 persons, while the town counted twenty separate households, having an average of 4.45 persons each. The precinct contained mostly single-person households, while the town had more families. Thirty-nine children under the age of 18 resided in the town, while only three lived in the precinct. All but one household in the town consisted of families, averaging 4.68 persons each. The skewed distribution of families helps to account for the lower average age in the town and the larger number of persons in each household compared to the precinct's demographic structure. Sixteen of the town families exhibited a nuclear structure, and only three were extended arrangements. In both the larger precinct and the town, US-born persons predominated over foreign-born inhabitants. The total population in the precinct and town of 147 consisted of 129 (88 percent) born in the United States and only eighteen (12 percent) from overseas. There were no Chinese present in either town or the rural portion of the precinct by 1910.

Of the thirty-two town inhabitants with listed occupations, twenty-two (69 percent) were mining-related and ten (31 percent) reflected service or commercial activities. The precinct population focused almost entirely on

Table 5.5. Average Age and Household Size, Susanville and Granite, 1870–1910

	1870	1880	1900	1910
Susanville				
Male age	32.02	37.18	27.14	35.52
Female age	21.88	24.4	20.1	23.70
Household size	2.9	2.8	2.8	2.45
Chinese age	33.60	33.63	46.6	52.47
Household size	7.5	8.0	3.5	1.6
Granite				
Male age	32.45	34.16	34.25	41.28
Female age	19.25	18.14	22.81	32.07
Household size	3.5	2.7	3.12	1.7
Chinese age	31.12	34.33	41.8	—
Household size	9.3	6.8	1.0	—
Granite Town				
Male age	—	—	29.26	31.77
Female age	—	—	22.69	23.11
Household size	—	—	2.91	4.45
Chinese age	—	—	31.33	—
Household size	—	—	*	—

**No separate households*

mining. Of the forty adults with occupations, thirty-four (85 percent) worked as miners, mine managers, or mining engineers. The other six included four forest rangers, one teamster, and one jeweler! In contrast to earlier times, the once bustling town of Granite's commerce, according to the census, had shrunk to just two general store merchants, one hotel keeper, one saloon proprietor, one postmaster, and one teacher.

The lack of systematic studies of the demography of western mining communities during the late nineteenth and early twentieth centuries made it difficult to establish what was typical or atypical about the evidence from eastern Oregon mining districts. Western mining historians have generalized that lode miners consisted of large numbers of foreign-born—for example, up to 40 percent of the Comstock mining community over time. California and Colorado showed similar results. As hard rock historian Mark Wyman put it, "The mining West became a conglomeration of nationalities."[81] The census records for Granite and Susanville, on the other hand, indicated that those districts never had more than 20 percent of foreign-born during the height of their mining operations. In 1900, Susanville's population was 18.6 percent

Table 5.6. Population of Select Mining Communities, 1900–1910

Precinct	1900			1910		
	Total	White	Chinese	Total	White	Chinese
Granite*	836	823	13	148	148	0
Granite Town	245	242	3	89	89	0
Susanville**	276	262	14	280	272	8

**Includes incorporated town. **Elk Creek precinct was renamed in 1890. Source: US Census Schedules*

foreign-born and Granite, 16 percent. By 1910, Susanville's foreign-born had dropped to 10 percent and Granite's, 12 percent.[82]

The gender composition in both eastern Oregon mining communities was almost identical and consistent over time. In 1900, Granite's males stood at 72 percent and females, 28 percent; Susanville's males came in at 70 percent and females, 30 percent. By 1910, the percentage of males and females was nearly identical: Each had about a 60/40 split between the two sexes. For comparison, the gender distribution for the Comstock showed a comparable breakdown in 1900, with males representing 53 percent and females, 47 percent, while by 1910, the number of males rose to 59 percent and females declined to 41 percent. Hopefully, future studies of other mining districts will provide the data needed to write with more confidence about the demographic experience of mining in the West.

VOL. XVIII.—No. 468. FEBRUARY 24, 1886. PRICE, TEN CENTS.

HOBSON'S CHOICE—YOU CAN GO, OR STAY.

February 24, 1886, cover of *Puck*, a popular 19th-century satirical magazine, showing a man wearing a hat labeled "Oregon" holding two guns and giving Chinese men a "Hobson's choice" or option of leaving by jumping off a cliff into the sea or staying and being shot to death. Courtesy of the Library of Congress.

Chapter Six
Chinese Miners in Eastern Oregon

The role of the Chinese in eastern Oregon mining is shrouded in myths, misconceptions, and racism. The Chinese arrived in Grant and Baker counties as early as 1866, according to newspaper accounts and mining records. Very soon, the Chinese immigrants represented a sizeable portion of the populations in the two counties (see Table 6.1) and remained a significant presence over the next twenty years. In the 1870 census for Grant County, 940 Chinese accounted for 42 percent of the total inhabitants and 69 percent of its miners; by 1880, the Chinese (905) had declined to 21 percent of the population, while the number of Chinese miners rose to about 82 percent of the total miners. The Baker County census data revealed a similar pattern over time: 680 Chinese represented 24 percent of the total population in 1870, but their numbers (787) declined to only 17 percent of the whole in 1880. The Baker County Chinese miners, however, increased from 53 percent of all miners in 1870 to 64 percent by 1880. By 1890, the Chinese numbers in Baker County had dropped to 6 percent of the total (398 out of 6,764), while their percentage of Grant County's population had also declined to 6 percent (326 out of 5,080). The reduction of Chinese inhabitants in both counties continued over the next decade, as reported in the federal census of 1900: Baker County Chinese accounted for 3 percent of the total (414 out of 15,597), while in Grant County, they made up 2 percent (114 out of 5,948).[1]

As noted in chapter three, the federal census for Grant County in 1870 and 1880 recorded that Chinese households contained, on average, ten to fifteen persons. At least some of these households were mining companies. In fact, in 1880, a census taker for Grant County separately listed nineteen

Chinese mining companies, ranging in size from four to twenty-four persons each and totaling 230 employees. One study of deed records for eastern Oregon found that between 1862 and 1900, Chinese purchases or leases accounted for 400 sales of mining claims. In 1870, William Packwood, an early mining entrepreneur in Baker County, wrote: "Chinamen have both bought claims at nominal prices and have paid as high as $35,000 for them."[2] Based on a variety of documentary sources, it seems fair to say that during the decade of the 1870s, eastern Oregon Chinese owned about 80 percent of all placer and hydraulic operations and accounted for a majority of the region's placer miners. The large numbers of Chinese miners in eastern Oregon were not unusual for the mining regions of the West. In both Montana and Idaho during the 1870s, the Chinese dominated placer mining, accounting for one-half to two-thirds of all such miners. As indicated above in chapters three and five, non-mining Chinese also lived in the mining camps of eastern Oregon. The 1870 and 1880 census reports, for example, revealed that the Chinese inhabitants in Susanville and Granite also included a few merchants, doctors, gamblers, cooks, tailors, and carpenters.[3]

The federal census records also revealed few female Chinese resided in the mining camps, especially after the passage of the Page Act in 1875. This Congressional legislation banned criminals, contract laborers, and "Mongolian" prostitutes from entry. As historian Mae Ngai has observed, "Chinese female immigration dropped precipitously—not because all Chinese women were prostitutes but because they were interrogated and inspected as though they were, and few were willing to subject themselves to the degrading treatment."[4] The population numbers for Chinese females in Grant County supported this point. The county had twelve Chinese females in 1870, eight in 1880, none in 1900, and one in 1910. The Elk Creek (Susanville) precinct listed two Chinese females in 1870, none in 1880 and 1900, and one in 1910; Granite precinct recorded one female in 1880 and none at any other decennial census.[5]

ATTITUDES TOWARDS THE CHINESE

The Chinese in eastern Oregon lived in a distinctly racist legal and social environment. The Oregon constitution of 1859 prohibited Chinese arriving after that date from owning real estate or mining claims. In 1862, the state legislature imposed an annual poll tax of $5 on all Chinese. Another statute in 1866 assessed a license fee of $4 per quarter on Chinese miners and $15 per month on all Chinese traders or merchants. The local county sheriff collected the fees or taxes. Most Chinese immigrants lived apart from their white neighbors, in either mining camps next to the placer claims or segregated urban districts referred to as "Chinatowns." Notable examples of the latter occurred at Baker City, Sumpter, Granite, Canyon City, and John Day (see Table 6.2). What we know about the Chinese experience in eastern Oregon comes chiefly from newspaper accounts and government records, such as mining claim filings, county court records, and federal mining and census reports.[6]

Certainly, the Euro American east Oregonians all too easily imbibed and reflected the anti-Chinese sentiments manifested throughout the Pacific Northwest during the last decades of the nineteenth century. Many white citizens had little patience for Chinese culture and lifestyle, especially opium smoking, and denigrated them as only temporary immigrants in the United States. The complaint about opium use may have been more about the "otherness" of communal smoking than concern with the drug itself. At the time, opium was legal and a common ingredient in patent medicines used by Euro Americans. Unfortunately, throughout the American West, negative attitudes led to periodic anti-Chinese riots in urban centers, discriminatory behavior, and restrictive laws.[7] An editorial in the *Grant County News* in 1885 reflected the negative sentiment held by some in Grant County:

> To every one it is apparent that the Chinese are a curse and a blight to this country, not only financially, but socially and morally; and but for their presence to-day the Pacific Coast would be peopled with a different class of laborers—white men, willing to settle down and make their home among us. What the Chinaman wears, he brings from China, and what he eats (except rats and lizards), he brings across the ocean, and thus American trade or production reaps no benefit from his presence.

The editor, however, assured his readers that he was not arguing for violence against the Chinese:

> But in order to rid the country of them we are not in favor of unlawful proceedings. The wrongs that white men suffer in consequence of the presence of the Chinese must be righted, not by mobs, but in accordance with law and order.

The editor then closed with a racist screed about the supposed ill effects of Chinese labor competing with white workers:

> [It] should [be] remember[ed] that the American laborer is directly affected by Chinese labor. Thousands of white men on the Pacific coast are suffering to-day because of the Chinese. Either

> they cannot get work at all, or if they can, it is at such rates as will scarcely pay their board. The Chinese not only crowd out white men, but white women also. They are in our kitchens, and they do all the laundry business. If a white widow woman would earn a living for herself and children by taking in washing, as hundreds do where there are no Chinese, she cannot do so because the almond-eyed, opium-smoking and rat-eating Mongolian is in the business, and has a monopoly of it.[8]

As late as 1892, the *Grant County News* editorialized that the Chinese were not desirable citizens:

> In discussing the merits and demerits of Chinese immigrations the other day one of our citizens said, in their behalf, that Oregon's industries are far advanced by reason of Chinese labor. The *News* admits that railroad building has been hurried somewhat on account of cheap heathen competition with honest white laborers, but contrast the degraded natives with the sturdy sons from Scotland, Germany and other nations which helped to people our land. They are here to build themselves homes and identify their interests with ours. Not so our yellow brothers.[9]

When the Chinatown in Canyon City burned down in 1885, its inhabitants were not allowed to rebuild and, instead, they moved two miles north to the town of John Day and became a part of that community. At this point, the John Day Chinatown was reputed to have almost 600 inhabitants.[10] Over time, the Chinese won a measure of toleration in John Day. As the 1902 history of eastern Oregon counties put it, "At present there are perhaps a hundred Chinamen in John Day. They have their own stores, three in number, and the community is apart from the main town. The inhabitants of this quaint settlement are orderly and apparently contented, and while, as is their way, they do not mingle with Americans nor do they adopt American manners and customs, they are not considered a detriment to the town." [11] Even when white observers later praised the Chinese as "efficient miners" and "very good workers," the recollections mostly emphasized their unusual customs and celebrations such as the Chinese New Year festivities. James Tabor, for example, noted in early-day Granite, "They [the Chinese] were very liberal on these days [New Years], giving the whites presents along with candy. Many firecrackers were set off."[12]

Regardless of the anti-Chinese views expressed in the press, Chinese miners, either individually or as a part of companies, actively plied their occupation in Grant and Baker counties. Despite the prohibition against owning mining claims, Euro Americans sold or leased to them such property, often involving significant amounts of money. In 1879, for example, the *Grant County News* noted that "the Chinamen have $100,000 invested in mining apparatus on the John Day River near this place."[13] At that time, about 600 Chinese were working placer claims over a ten-mile stretch of the river and had paid a total of $6,000 for just three of the claims within the work area. According to the Grant County deed records, between 1866 and 1894, Chinese mining companies made most of their purchases from Euro American claim holders. Typically, the sales and leases ranged from $50 to $4,000 and usually included equipment such as picks, shovels, pans, and rockers.[14] Intriguingly, an advertisement in the *Grant County News* gave notice that five Chinese men had bought the ranch of Charles Staley for $1,400 in October 1880. [15] Unstated was whether this involved mining or other purposes. It should also be noted that Euro American mine owners, such as those operating the Monumental, Connor Creek, and Virtue quartz mines, also occasionally hired Chinese labor. As William Packwood observed in 1872, "Chinamen are in the country and can be hired cheaply," usually for half the wages paid to white men.[16]

It did not always go well when white miners were expected to work next to the Chinese in a lode mine. The owners of the Monumental Mine, for example, had to assure their Euro American employees that the Chinese workers had been hired only for the season and would be doing only "a large amount of surface work . . . of such a nature that the only laborers that could be found to do it were Chinamen and . . . [would be kept no] longer than the work was done."[17] The hiring of Chinese at the Cabell Brothers' La Belleview Mine near Granite proved even more unsettling. The white miners expressed their indignation by dynamiting the Chinese bunkhouse. In its January 31, 1889, issue, *The Grant County News* reported:

> The Chinamen were in their cabin when the explosion occurred and the whole top of the house was lifted in the air as if by a Kansas cyclone. One or two of the Mongolians were dangerously injured and others slightly. How the explosion occurred is not known, but we understand that it is supposed to be the work of white men who are incensed at the action of the Cabell Bros. in employing Chinese miners in preference to white men and they take this means of retaliating, a very poor method indeed, if it is true. The Chinese are not to be blamed for accepting work even if in direct competition with white men. The Cabell Bros. are the parties to be blamed, but in either case it is wrong to resort to such

> means to remedy the matter and the perpetrators of the outrage should be severely punished if apprehended.

As described above in chapter three, the Virtue Mine outside of Baker City was unusual in hiring Chinese as underground laborers. Except for California, throughout the West, mine owners and their white workers generally excluded the Chinese from working in lode mines.[18]

Disputes between Chinese and white miners did not always end in the favor of the whites. A humorous article in the *East Oregonian* recounted one such struggle on Granite Creek in May 1877. According to a correspondent of the newspaper,

> The peace and quietude of the town of Independence [Granite] was disturbed on the morning of the 5th inst., over a report that war had been declared between the Mongolians and certain representatives of the white population residing on Granite Creek. . . . The cause of the trouble . . . was a dispute over water rights which both parties claimed, and which neither were disposed to relinquish. Diplomacy could accomplish nothing. War was inevitable."

In the ensuing struggle, the white miners, under two leaders, fortified the divide between Granite and Last Chance Creeks. Their position seemed impregnable, but a large number of Chinese miners outflanked the white redoubt and in the following melee, captured one white leader, sending his forces into retreat. The remainder of the whites, under the other leader, "surrendered unconditionally to the moon-eyes." The author of the piece reported that for all their efforts, the white miners appeared "defeated, worn out and humiliated." Subsequently, both sides held a council of peace and resolved their differences, with the correspondent writing that "it is generally thought there will be no further trouble."[19]

The Chinese in eastern Oregon also established mercantile businesses to support the Chinese miners and others in the region. These companies especially supplied labor contracting, medical, and social services to the Chinese community. The stores also had broader economic power within the greater region where they operated. For example, in 1879, the *Grant County News* estimated the "number of Chinese stores in the vicinity, 10, of which Kam Wah Chung and Company carry about $100,000 in goods; Hong Chong and Company carry $30,000."[20] US Census records indicated that Chinese merchants operated continuously in both Granite and Susanville throughout the mining era. Both Euro Americans and Chinese in Baker and Grant counties

Kam Wah Chung store in John Day on the left, with a Jos House Shrine in the center at the end of the street. Courtesy of the Grant County Museum.

patronized the Chinese stores, expanding their economic impact. The *Blue Mountain Eagle* noted in its October 4, 1901, issue that the Sing Lee Kee & Co at Susanville had "a large patronage." Ing Hay and Lung On, proprietors of the Kam Wah Chung Company from 1887 to the 1940s in John Day, provided goods and services to Euro Americans long after most Chinese had left Grant County. Ing Hay practiced pulsology to diagnose ailments, which he then treated with Chinese herbal medicine, while Lung On carried on the mercantile business and made other successful investments.[21]

As the Chinese population dwindled in the 1890s and early 1900s, the public perception, as recorded in newspapers and other publications, of the Chinese immigrant miners changed. Summing up this shifting view, archaeologist Don Hann has perceptively written that "those depictions were no longer of workers organized into mining companies but of laborers working for low wages in white-owned mines or independently scraping out a meager existence by rewashing the waste rocks left in abandoned mines to recover scraps of gold."[22] An early example of the tendency to denigrate the Chinese mining practices and, by extension, the economic impact of their labor on the region, appeared in a 1902 history of eastern Oregon counties. It stated that "during the latter 'sixties' the population here began to scatter, the placer mines giving out, and all through the 'seventies' the Chinese were about the only miners who considered the tailings on Canyon creek worth re-working."[23] Non-mining Chinese members of eastern Oregon communities could gain acceptance, albeit grudgingly. In an article about Susanville in 1901, the *Blue Mountain Eagle*, referred to a "shrewd," multitalented Chinese man named Ti. According to the *Eagle*,

> Ti is at once a blacksmith, cook, laundryman, race horse trainer, watch repairer, placer miner, barber, shoemaker, horse breaker, and an all around mechanic. He can shoe a horse or fry a flap-jack with equal ease, 'do-up' a shirt or train a trotter, repair your watch or pan out gold, shave a man or mend his shoes, break the wildest horse in camp or perform any mechanical feat known to Susanville's needs. Miners have no love for the Chinese, and as a rule the people of Susanville are not abnormal in this regard, but no one has a word to say against this genius of the celestial kingdom. Ti is a shrewd looking chap and in his varied pursuits needs every whit of his ability.[24]

Reflecting the emergence of a denigrating view of the Chinese miner, Rodman Paul, a leading historian of western mining, remarked about Oregon mining that "as the placer yield declined . . . the Chinese, those harbingers of decline, became numerous" in the 1870s.[25] As late as the 1970s, a historian could write that "when the easy riches of the placer mining process and the quick returns from the hydraulic and the dredge operations were exhausted, the Chinese miner came into his own. The remaining method of mining was the patient, difficult process of panning the great gravel banks left by the high-pressure hoses and the dredges. Because this work was laborious and did not offer quick returns, many white miners tended not to take it up. They quickly sold or leased their claims to Chinese miners."[26]

William Trimble, author of the 1914 classic, *The Mining Advance into the Inland Empire*, also emphasized the secondary role of the Chinese miner when discussing the wastefulness of placer mining: "Waste was to a considerable degree relieved by the incoming of the patient Chinese into every camp after the cream had been taken by the whites."[27] The notion that Chinese miners only reworked the mine tailings and worn-out ground left by white miners was reinforced in numerous pioneer memoirs and reminiscences published in books and newspapers during the first half of the twentieth century. Unfortunately, as Hann has noted, "Depictions of the Chinese miners' being marginally successful primarily through hard work and persistence, rather than through expert application of mining techniques and managerial skill, continue to be made in both academic and popular publications."[28]

CHINESE MINING TECHNOLOGY

Recent historical and archeological scholarship has emphasized that the Chinese miners of eastern Oregon were well-organized, technologically savvy placer and hydraulic miners. Most Chinese immigrants to western North America in the nineteenth century came from three counties in Guangdong Province that had long-term mining experience not only in China but also in the gold mines of Borneo. These miners worked solo, with partners, or formed egalitarian cooperative ventures called *kongsi*. The members of the cooperative organizations did not work for wages, but split profits based on the percentage of capital, goods, and/or expertise brought to the enterprise. All the partners got to vote when selecting the company leadership positions and making major decisions. The Chinese used this approach in medium or smaller placer mines. In the case of large mining operations, a wealthy owner would supply expensive hydraulic mining apparatus and hire wage laborers to do the work. The personnel of the Chinese mining *kongsi* typically included the head of the operation, at least one cook, a gardener, a woodworker, a blacksmith, foremen, general laborers, and a clerk for record keeping and store management. The strength of the organization came from the focus of the partners on common goals and the ability to unite the group against external threats.[29]

The placer mining techniques developed by the Chinese prior to arriving in the North American gold fields worked well in eastern Oregon. Chinese

A typical scene of Chinese placer miners in California in the 1850s. Courtesy of the Bancroft Library.

miners in eastern Oregon were adept at creating extensive ditch and reservoir networks to capture and transport water from distant sources to locations above placer mines and use that water to wash away the surface sediment and expose the gold-bearing deposits. Chinese laborers built the large-scale Sparta and Eldorado Ditch networks between 1864 and 1873. The Eldorado Ditch ultimately extended more than 100 miles and was 8.5 feet wide at the top, 6 feet wide at the bottom, and 3 feet deep. As many as a thousand Chinese may have worked on the Eldorado Ditch, and 300 labored on the Sparta Ditch. William Packwood, an early mining and business entrepreneur in Baker County and the main force behind both ditches, had a close relationship with the Chinese labor contractors who supplied the workers. In particular, the application of extensive hand-dug ditch networks to carry water to the placer grounds proved crucial for successful mining. The transported water allowed miners to remove efficiently the sediments on top of the gold deposits and to separate the gold from the surrounding matrix composed of soil and gravel by applying sluicing methods. Both Chinese and Euro American miners also employed hydraulic technology, which used manufactured metal nozzles fed by pressurized water from metal pipes (penstock) to remove the overburden, enabling them to process the gold-bearing debris.[30]

Documentary and physical evidence indicated the presence in eastern Oregon of *kongsi* mining partnerships. Census records, newspaper references and mining claim deeds named Chinese-owned mining companies clearly operating as partnerships. In addition, as Hann has written, "Archaeological evidence demonstrates that food and other supplies were imported from China for use in the mines and that the Chinese miners shared resources, the remains of which are found tightly clustered in mining sites. The Chinese structures documented at many sites may have served as the *kongsi* central office, warehouse, and cookhouse. Trade specialization, including cooks, gardeners, and blacksmiths, is evident in the archaeological remains recovered at the mining camps."[31]

As noted above, the federal census for 1870 and 1880 recorded several households in Grant County with ten or more Chinese living together. Some of these households probably were *kongsi* headquarters, providing temporary living areas and storehouses for the partners near their mining operations. The 1870 federal census for "Products of Industry" recorded three Chinese companies operating at Susanville: Sam-Chong and Gee employed twelve miners each, and Ah Fat employed six. The 1880 census taker listed the nineteen Chinese mining companies by name in Grant County, averaging a dozen or so men each. As late as 1900 and 1901, the *Sumpter Miner* noted a Chinese mining company operating at Susanville. The archives of the Kam Wah Chung contained leases involving four different Chinese companies, two

A caricature of the Chinese miners' lifestyle in California during the 1850s. Courtesy of the Bancroft Library.

of which were referred to as co-partnerships. The parties entered the leases between 1887 and 1898. Most of the mining transactions listed in the Grant County deed records specifically denoted the Chinese party as a company, not as individuals. "The consistency of these documentary records," Hann argues, "indicates widespread, contemporary recognition of the company framework for Chinese mining in the region."[32] Indeed, as the compiler of *The Resources of the State of Oregon* issued by the State Board of Agriculture in 1899 observed, most of the placer mines in eastern Oregon were "generally owned or worked by companies or associations of whites or Chinese, the tendency being to consolidate the mining interests."[33]

ARCHEOLOGICAL STUDY OF CHINESE MINING CAMPS

Recent archeological field investigations in eastern Oregon have begun to reveal the lived experience of the Chinese miners employing the technological and business aspects of the *kongsi* partnership arrangement. The size of placer mining sites could be quite extensive, covering several hundred acres. Within that area, the miners placed their habitation sites near the active hydraulic operations. Artifacts, though, usually were clustered in discrete locations, whether household or mining implements. Domestic items included cookware, dishes, utensils, food containers, as well as fragments of clothing and boots. Mining objects included hand tools and a variety of hydraulic equipment, such as fragments of metal piping and nozzles. Many artifacts came from China, indicating a supply network across the ocean.[34]

The artifact concentrations were often found in locations containing the foundations of habitations. Archaeologists have interpreted these as support for log cabins or tent platforms. These could have been built by the Chinese or taken over after buying or leasing a claim, as these items were usually part of the agreement. Archaeologists found U-shaped, dry-stacked stone structures as an intriguing feature of the Chinese mining sites. Excavations showed such features to have been used as hearths and associated with foundations for cabins or tents. The archaeologists speculated that the larger of these structures may have served as a headquarters for the *kongsi*'s key activities related to managing their operations. In addition, most of the mining sites provided evidence of metalworking. For example, archaeologists have documented a Chinese blacksmith shop at the Ah Heng site. Such labor kept equipment in repair or repurposed metal cans and broken tools to perform a variety of food preparation, household, and mining functions. Archaeologists have also found remnants of "handmade nozzles for low-pressure hydraulicing made from riveted sheet metal . . . and demonstrate the skillful adaption of new technology."[35]

Excavated mining sites indicated centralized food preparation and cooking. Cookware, crockery, and storage containers reflected both China and Euro American origins. The Chinese used stoneware, referred to as "Chinese Brown Glazed Stoneware," to preserve food, sauces, and liquor imported from China. Remnants of ceramics exhibited standardized designs typical of imported Chinese products from Guangdong and Jiungxi provinces. Recovered faunal remains and bones demonstrated that the Chinese miners made use of the surrounding plants and animals to supplement foods from their own garden plots where they raised vegetables.

Detailed excavation and analysis of the stacked rock features at the Ah Hee Diggings site near Granite and Happy Camp and Ah Heng sites not far from Susanville have helped to reveal "evidence of how residents spent their time, what they purchased, what they ate, and how they organized their living spaces."[36] Findings from these eastern Oregon mining sites, combined with field work from locations in California and Montana, as well as the goldfields of Australia and New Zealand, provided evidence of the common elements of the Chinese wherever they mined. The ubiquitous stacked rock features, widespread artifact assemblages, and similar spatial-use patterns appeared at all these archeological sites related to mining. The archeological investigations revealed that the Chinese mining camps contained specific areas for living and carrying on support activities adjacent to the placer deposits being mined.

The demographic data from the US Census reporting on the Chinese population at Granite and Elk Creek (Susanville) presented in chapter three complemented the archaeological evidence recovered at those two locations. In both places during the 1870s and 1880s, the Chinese miners dominated the communities. They accounted for 81 percent (365) of the Granite population in 1870, declining to 61 percent (123) by 1880. In Elk Creek, on the other hand, the Chinese count went from 58 percent (60) in 1870 to 85 percent (128) of the inhabitants in 1880. In both communities, the census records revealed that the Chinese overwhelmingly worked as miners—90 to 98 percent—but other occupations were listed. At Granite in 1870, for example, the non-miner Chinese included a doctor, twelve cooks, eleven gaming house operators, and nine merchants. By 1880, the only non-mining Chinese at Granite included two cooks, one storekeeper, and one house servant. Over both censuses, the average Chinese household size averaged eight to nine, a few reaching as many as twelve persons or more. Again, in Elk Creek, mining accounted for most of the Chinese occupations listed. There, in 1870, the census taker recorded fifty-six miners, two Chinese gamblers, and two females (no occupations given), while in 1880, the non-miner Chinese included five cooks and one carpenter. The Chinese household size in Elk Creek mirrored that in Granite. The large Chinese households in both communities reflected the organizational structure of their mining operations.

The mining camps, however, did not exist in isolation. Chinatowns in John Day, Baker City, Sumpter, and Granite functioned as mercantile, labor contracting, medical support, and social centers for these outlying settlements. When the six-month placer mining season was over, the Chinese miners would move back to the Chinatowns to live. The documents, archive, and physical objects at the Kam Wah Chung State Heritage site in John Day demonstrated the connection between the mining camps and an urban Chinese setting. Goods found at the mining sites could be purchased from the Kam

John Day, Oregon, ca. 1910. Main Street is in the center of the foreground, and Chinatown is to the right of the large schoolhouse in the upper center of the photograph. Courtesy of the Grant County Museum.

Wah Chung Company; and the archived business records, such as lease and purchase agreements and other business receipts, indicated that Kam Wah Chung also acted as a records repository for the *kongsi* mining companies and may have been partners or investors as well. The Kam Wah Chung partners, Lung On and Ing Hay provided multiple services for the Chinese community in Grant County. In addition to the medical and general merchandise side of their business, their store building housed a religious shrine where visitors could pray and honor their family deities. The two men also furnished letter writing and interpretation services to the local Chinese. Also, the business served as an employment and general information center. Their merchandise included both Chinese and Euro American products. Over time, they became respected and well-liked fixtures in the greater John Day community until their deaths in the middle of the twentieth century. Research in the written records and objects held by the Kam Wah Chung Museum is the source of ongoing historical activity.[37]

ESTIMATES OF CHINESE MINING PRODUCTIVITY

It is difficult to determine how much gold the Chinese miners recovered in eastern Oregon in the late nineteenth century, but it must have been considerable. They were adept at concealing what they mined, sending much of it back to China instead of selling it to the US Mint in San Francisco.

In 1883, the frustrated director of the Mint reported, "It has been difficult to obtain reliable information as to the amount of gold and silver annually produced from the mines in Oregon, for much of the mining has been done by the Chinese, who are reticent as to their operations, and much of the dust obtained from placer mining in this State is sent by private mail or carried to San Francisco or sent out of the country by the miner himself."[38] The personal return to China with gold had its own dangers. The *Grant County News*, for example, reported in 1886 that three Chinese men, headed back to China, were robbed of $1,500 during a holdup of the stage running between Canyon City and Baker City.[39]

A report written a decade earlier (1871) by the federal commissioner of mining statistics expressed similar sentiments about the difficulty of establishing Chinese mining returns in eastern Oregon:

> The influx of Chinese, who succeed, by purchase in most cases, to the claims formerly worked by the white, and who, by their superior patience and economy, continue the production of gold in many localities where it would otherwise cease. It is very difficult, however, to ascertain the amount of production from such sources.

The commissioner then went on to give an example of his predicament:

> The reports from sixty-four placer-claims in Grant County, eleven of which are worked by white men with paid labor, and the remainder by Chinese owners, show for the former a yield of $4 per day per hand, and for the latter only $1.30. There is no doubt that the Chinese have in this case concealed the actual amount of their production, reporting an aggregate of about $126,000, when the true amount must have been at least twice as great.[40]

Scattered references in the contemporary press and other publications reinforced the popular perception that the Chinese miners hid the results of their efforts. A few publications, however, did provide scraps of information about the returns from Chinese mining. For example, a promotional book about the resources of the Pacific Northwest produced by the *Oregonian* in 1894, noted that "the Chinese were especially fortunate in Eastern Oregon during 1892, they having secured during that year about $150,000 in placer gold."[41] This accounted for a little more than 14 percent of the total gold and silver mined in eastern Oregon that year. The Baker City *Morning Democrat Souvenir Edition* (1898) claimed that in the Sparta District of Baker County

Table 6.1. Chinese in Oregon, Baker, and Grant Counties, 1870–1910

	Oregon	Baker County	Chinese Miners	Grant County	Chinese Miners
1870	3,330	680	591	940	855
1880	9,510	787	638	905	790
1890	9,540	398	—	326	—
1900	10,397	414	101	114	38
1910	7,363	90	11	37	4

Source: US Census Schedules

during the early 1880s, "Three Chinese companies . . . used the mail . . . and sent to San Francisco weekly for two years fully $4,000.00."[42] A promotional pamphlet touting the placer riches near Sumpter in 1900 reported that "a lot of Chinamen have, for months past, been washing out gold enough . . . to pay $320 a week in royalties."[43]

More typical, though, was a February 27, 1901, article in the *Sumpter Miner* describing the early days of mining at Sumpter. It stated that "soon the Chinese came pouring in. They picked about over the choicest ground unmolested for over thirty years, and the amount of gold these Mongolians sent to China will never be known, although it is estimated at many millions." The *Miner* repeated this refrain three years later, stating "no one know how much gold the Chinese have mined."[44] The anonymous author of the 1902 history of eastern Oregon counties, also noted that the Chinese "invariably decline to give definite information as to the amount of their [mining] product."[45] Reflecting the stereotypical attitude of the times, the author, in referring to the Chinese miners at Sumpter, commented that "no estimate can be made of the amount of gold which rewards their labors, as inquiry into this delicate matter of private business invariably elicits the same reply: 'Some days belly well, some days no good at all'."[46]

Wherever they worked, the Chinese experienced discrimination and sometimes violence. Although outside of the immediate geographic focus of this study, the horrific massacre of thirty-four Chinese miners near the confluence of the Snake River and Deep Creek in northeast Oregon's Wallowa County in 1887 was a case in point. In an atrocity that made national newspaper headlines, a gang of six or seven Euro Americans dressed as Native Americans attacked, scalped, and beheaded the Chinese miners for their gold. Authorities captured six of the perpetrators and tried them, but a jury found them not guilty in August 1888. Shortly after the trial, all the court records disappeared until rediscovered in 1995. Historian Sue Fawn Chung aptly commented:

> This event demonstrated how dangerous life could be for Chinese miners, how many deaths may never be discovered, how justice

> was not on the side of the Chinese, and why the Chinese mined in groups that shared the common bond of native place . . .— because mutual trust and protection were vital in placer mining sites even though safety was uncertain.[47]

Table 6.2. Population of Baker and Grant County Chinatowns, 1870–1910

	Baker City	John Day	Canyon City	Sumpter
1870	29	85	162	–
1880	166	357	79	–
1890	121	600 *est.*	–	–
1900	264	31	9	32
1910	37	7	2	14

Source: US Census Schedules

Studies by historians Sue Fawn Chung and Mae Ngai of Chinese miners and merchants in selected communities across the American West and internationally record a similar story to that of eastern Oregon. In Baker and Grant counties, and especially the communities of Susanville and Granite, the Chinese brought an expertise in placer mining that accounted for a major portion of the non-lode gold produced in the late nineteenth century. In addition, the federal census records track the rise and then decline in the Chinese presence in the region. As placer mining receded after 1900, the main Chinese population shifted to urban centers such as Portland and Astoria, taking up various lines of work. Their new occupations included cannery work, truck gardening, laundryman, and cook, to name a few. The census records for Grant County, however, reveal that a few elderly miners and determined merchants persisted at least until 1910 and even beyond.[48] Between 1900 and 1910, most of the remaining Chinese in Grant County worked as farm laborers, herders, or cooks.[49] The historical and archeological record has demonstrated a similar pattern of living and laboring among Chinese miners and other workers, whether in eastern Oregon, Nevada, Idaho, Montana, northern California, British Columbia, or Australia.

The above-ground physical reminders of the Chinese in eastern Oregon are few: The El Dorado and Sparta Ditches, Ah Hee Chinese Rock Wall, the Chinese cemetery in Baker City, and the Kam Wah Chung and Company building in John Day stand out. Fortunately, ongoing archaeological research of Chinese mining sites also helps to fill in the gaps in our knowledge of the Chinese contribution to the building of eastern Oregon. Continuing documentary research in the Kam Wah Chung archives and other sources may provide further evidence of the role their mining operations played in the region's development.

The third and most successful of the Sumpter dredges. It operated between 1935 and 1954. Courtesy of the Baker County Library.

Chapter Seven
The Decline in Oregon Gold Mining

In the late nineteenth century, the monetary policy of the United States became a political lightning rod as various agrarian and labor movements confronted the modernizing effects of industrialism and urbanism. The political results of this confrontation played out both at the national and state levels. The reformed-minded challenge of the Populists to the dominant Republican and Democratic parties attracted pockets of support in Oregon during the 1890s and found a strong following in Baker and Grant counties. Although the Oregon Populists never displaced the state's major parties, their efforts did lead to significant political reforms in the early twentieth century known as the "Oregon System." While Oregon's turn-of-the-century politics became more progressive, the state's economy simultaneously experienced growth and prosperity. During a time of great change after 1900, gold and silver mining struggled to find its footing as Oregon's timbermen and agriculturalists focused on exploiting its forest and field natural resources. Even with the rise in the price of gold over the course of the twentieth century, other factors worked against the return of large-scale lode mining.

MINING AND MONETARY POLITICS IN THE 1890s

The profitability of precious metal mining was hostage to both the monetary policy of the US government and the global production of gold and silver. North American mine owners could attempt to influence US monetary policy but had little control over global mining operations. In the second half of the nineteenth century, the US government's monetary decisions created a deflationary economy that caused financial hardship for many debtors, farmers, and small businessmen as they had to repay fixed-dollar obligations like mortgages and bank loans with lower revenues as prices for commodities and other goods fell. In addition, American mining interests suffered lower financial returns for gold and silver as new precious metal discoveries in Alaska, South Africa, and Mexico put downward pressure on the prices for these commodities. The world production of gold, for example, increased from 10–20 tons annually between 1800 and 1850 to 450 tons per year in 1900.[1]

In 1791, Secretary of the Treasury Alexander Hamilton recommended, and Congress approved, placing the United States on the bimetallic standard, making both gold and silver legal tender. He set the value of gold at $19.39 per ounce and silver at $1.29 per ounce—a ratio of 15 to 1. In 1837, Congress adjusted the mint price to $20.67 for gold, while silver stayed at $1.29 (a ratio of 16 to 1). The two metals remained at those valuations until 1873, when the United States moved to the gold standard and demonetized silver, except for small change worth less than $5 (dimes, quarters, and half-dollars). The United States's adoption of the gold standard reflected a trend occurring throughout Europe in the 1870s, initiated by Great Britain.[2]

The move to the gold standard supported the price of gold and led to the selling of silver, which, coupled with the heavy output from the Comstock Lode and other silver-producing locations, resulted in the 1890s with silver valued at half its official price—roughly 63 cents per ounce. Pro-silver forces fought back with the Bland-Allison Act of 1878 and the Sherman Silver Purchase Act of 1890. These measures required large monthly purchases of silver by the US Treasury at $1.29 per ounce, but neither act brought back the free and unlimited coinage of silver. Most of the silver minted by the government never circulated, as the public preferred the convenience of a paper currency backed by gold reserves. The main beneficiaries of the government's purchase of silver were western mines. By 1885, the mint had coined about two hundred million silver dollars, but only one-quarter of that circulated.[3]

The gold standard limited the money supply in a growing economy, which resulted in a price deflation benefiting creditors and hurting debtors. By one estimate, prices fell about 1 percent a year between 1865 and 1897. As historian Richard White observed, "Wealthy creditors gained premiums beyond interest payments since deflation meant that the dollars paid to them in interest, and ultimately in repayment of principal, were always more valuable than the earlier dollars they had lent. The monetary system transferred wealth from the debtor West and South to the East, whose banks and investors controlled the money that was lent."[4] The stringent controls on international trade and currencies imposed by the adherence to the gold standard led to a banking crisis and the Depression of 1893, causing great hardship in much of the nation for the rest of the decade.[5]

In the last third of the nineteenth century, Oregonians did not escape the broader effects of a deflationary economy. Markets for Oregon products from field, forest, and mining were distant; and high transportation costs and competition from better located producers combined with limited investment funds to make economic progress difficult for many Oregonians. In particular, wheat farmers, sheep ranchers, and miners all found themselves at the mercy of commodity prices subject to world market conditions and national tariff policies that favored consumers and industrial interests over producers. Disenchanted farmers, livestock raisers, and hard rock miners sought relief from their plight by engaging in radical politics. In the 1880s, large numbers of Oregon farmers joined the Patrons of Husbandry (Grange Movement) and the Farmers' Alliance to lobby for laws enabling the collection and publication of agricultural statistics, regulation of railroad rates and monopolistic corporations, support for the Oregon Agricultural College in Corvallis, and the development of experimental farms to develop improvement in crops, livestock breeding, and fertilizers. Wage-earning miners working underground in lode mines formed unions, such as the Western Federation of Miners, to seek better pay and working conditions. By the early 1890s, many agrarians, miners, and others seeking specific reforms—such as the prohibition of alcohol, women's suffrage, the single tax (which sought to end poverty by taxing the inflationary value of real estate), and the free and unlimited coinage of silver—coalesced into the People's Party.[6]

The People's Party touted an agenda of reform calling for graduated income tax; government ownership of railroads, telegraph and telephone companies; limits on immigration; free coinage of silver; the secret ballot; and direct election of senators; and a subtreasury system to support farm commodity prices. The demand for the unlimited coinage of silver especially attracted the support of western mining interests and agrarians for its inflationary effects on the nation's economy. In eastern Oregon, the Populists in both Baker and Grant Counties started newspapers to spread their political ideas. The *Epigram* came out in Baker City while the *Sentinel* held forth from the town of John Day; both papers lasted for most of the 1890s.[7]

While Oregon's Republican and Democratic parties continued to dominate public offices in the 1890s, a few Populists won election to the state legislature and county offices and polled well in the election for United States representatives and presidential candidates. Many members of both the Republican and Democratic parties favored silver, and the formation of the Populist Party gave them an alternative to the political status quo. The anonymous author of the *Illustrated History of Baker, Grant . . . Counties* wrote, "[In] Baker County especially were conditions favorable for the success of the Silver movement, as the placing of the white metal on an equal footing with gold would open up the silver mines in the western part of Baker county and those of Grant county."[8]

In November 1892, the Populists of Baker County held their first convention, and a few months later, forty prominent miners, ranchers, merchants, and bankers organized the "First Free Coinage Silver Club of Oregon" under the banner of "Equal rights for gold and silver, free coinage for both." The results of the June 1892 elections gave the Populists one county-wide office: assessor. By 1894, their message had gained increasing support. That year the Baker County Republican convention, on the defensive, found it necessary to adopt a free silver plank. The Populists, however, went on to win seven of twelve county offices on the ballot, while their candidate for congressman carried the county. At the state level, the Republican party convention remained silent on the silver question and failed to embrace openly the gold standard. The election of 1896 was the high point for populism in Baker County. William Bryan, the Populist candidate for president, received 1,870 votes while William McKinley received only 957. Baker County, in fact, was the banner silver county of the state that year. At the state and local elections in June, the Populists held on to seven of twelve county offices and helped send a Populist representative to the state legislature. Success proved fleeting, however, as the Populists statewide were subsumed into the Democratic Party by 1898.[9]

In Grant County, the Populist Party organized in tandem with Baker County in the spring of 1892. The Populists, however, never fared as well at the polls as in the neighboring Baker County. The Republicans generally outperformed both the Democrats and the Populists, mainly because the large livestock sector of the county's economy preferred the high protective tariff plank of the Republican Party. This factor outweighed the county's mining interests desire for a reform of the nation's monetary policy in favor of the free coinage of silver. In 1896, Bryan did, however, carry Grant County over McKinley, receiving 867 votes to 738 for McKinley. Grant County also comprised part of the vote that sent a Populist representative from eastern Oregon to the state legislature.[10]

During the first half of the 1890s, the Populists' insurgency at the state level achieved a respectable showing at the ballot box but never enough to fully overturn the established Republican and Democratic parties. In the election of 1892, for instance, James Weaver, the Populist candidate for president, garnered 34 percent of the vote while Benjamin Harrison (Republican) got 45 percent, and Grover Cleveland (Democrat) won only 18 percent. Oregon ranked tenth among the states giving the Populists the largest presidential vote totals. Weaver did well in eastern Oregon, getting 36 percent of the vote in Baker County and 22 percent in Grant County. Weaver received one electoral vote out of Oregon's four cast that year. The Populist candidate for Oregon's

Second District Representative (covering eastern Oregon), John Luce, did not fare as well in his race, receiving only 17 percent against 45 percent for the Republican, William Ellis. The Democrat, James Slater, came in second with 38 percent. In the 1894 election, the Populist candidate for the Second District Representative did marginally better (getting 27 percent) against Ellis, who won reelection with 48 percent of the vote. The Populists elected five members of the lower house of the state legislature in 1893, ten in 1895, and thirteen in 1897.

In the statewide election of 1896, William Jennings Bryan, the Democratic/Populist candidate for president, narrowly lost to the Republican William McKinley, 48 percent to 50 percent. Baker County gave McKinley 65 percent to Bryan's 33 percent while Grant County voted 52 percent to 45 percent for McKinley over Bryan. The Democratic/Populist candidate for United States Representative in eastern Oregon came much closer to winning his contest. Martin Quinn garnered 29 percent of the vote, while Republican Ellis squeaked by with 30 percent. Three other minor party candidates split the remaining votes cast.

Oregon's state-wide political scene in the 1890s was scrambled by a split in the Republican Party over whether to remain on the gold standard or switch to the free and unlimited coinage of silver. One faction led by Senator John H. Mitchell ultimately came out for the gold standard while banker Henry Corbett, lawyer Joseph Simon, and wealthy mine owner Jonathan Bourne—once a supporter but now an opponent of Mitchell—saw an opportunity to dump Mitchell by aligning with William U'Ren, a leading reformer and Populist. Since Mitchell was up for reelection in 1897, the moment for both men had arrived. U'Ren and twelve other Populists had won elections to the legislature and held the balance of power in the choice of Oregon's senator. Bourne agreed to support U'Ren's package of reform measures; and through a complicated and questionable series of actions that prevented the legislative session from organizing, they stymied Mitchell's reelection and then pushed through the reform acts in the next meeting of the legislature—chiefly the initiative and referendum amendments to the state constitution. Some political observers at the time believed that corruption played a key role in the process, and Bourne later privately admitted that money and other favors did change hands during the 1897 "hold-up" legislative session. Some evidence even suggests that Bryan's loss to McKinley by less than 2,000 votes was the result of voter fraud.[11]

ECONOMICS OF PRECIOUS METAL MINING SINCE 1911

While 1910 marked a sharp drop in gold and silver production from the boom years of 1900 to 1905, the mining of precious metals continued in eastern

Oregon until the eve of World War II. As Table 7.1 shows, except for the decade of the 1920s, the mines of Baker and Grant counties averaged at least a million dollars a year. The average number of active placer and lode mines in the two counties over the forty-year period certainly suggests a lot of mining activity: 1911–1920: 22; 1921–1930: 27; and 1931–1940: 51. In fact, just a few mining operations accounted for most of the gold and silver produced during those years. Although attempts at mechanical dredging in eastern Oregon began as early as 1898 near John Day, a successful large-scale bucket line dredge began working in the Sumpter Valley of Baker County only in 1913. Between 1913 and 1954, the Sumpter dredge output amounted to $12,250,000 recovered from 2,603 acres of land in the Sumpter Valley. In all, 60 million yards of gravel produced almost 270,000 ounces of gold. Additionally, dredges in the John Day River in the vicinity of John Day and Prairie City produced about $5 million in gold and silver between 1916 and 1949, while smaller recoveries by dredging occurred along the Middle Fork of the John Day River and on streams in the vicinity of Granite. From 1913 to 1940, two lode mines in the Cornucopia district and one in the Mormon Basin district accounted for most of the underground yield during the period. The Cornucopia mines generated over $10 million, and the Rainbow mine in the Mormon Basin, about $2 million.[12]

In the 1920s, mining suffered from a lack of investment during a booming economy, with the stock market seeming to offer better returns than speculative mining proposals. Post-World War I inflation also raised the cost of labor and materials, directly increasing mining expenses. The Depression of the 1930s, on the other hand, changed the financial dynamics of mining precious materials, giving a new lease on life, at least in the short term. Hard times cut wages and the cost of equipment and supplies, making mining once again a more attractive employment and investment opportunity. The major boost for reviving mining came from the increase in the price of gold. In January 1934, President Roosevelt raised the price from $20.67 to $35.00 an ounce. The increase in price led to a peak output for gold for both the United States and Oregon in 1940: $171.5 million and $3.9 million, respectively. The Oregon production in 1940 came from 112 lode and 192 placer

Table 7.1. Oregon Gold and Silver Product, 1911–1940

Years	Oregon	Baker and Grant Counties
1911–1920	$14,027,765	$11,939,025
1921–1930	$4,713,289	$3,521,968
1931–1940	$18,912,430	$12,798,435

Source: Brooks and Ramp

mines, with Baker and Grant counties accounting for forty-three of the lode and sixty of the placer mines.

At the end of the 1930s, however, as the United States began to pull out of the Depression and World War II began, wages went up and materials became scarce and more costly, putting renewed pressure on mining. Then, in 1942, President Roosevelt issued the War Production Board Order L-208, which stopped gold mining, diverting manpower and equipment to mines producing metals and minerals essential for the war effort. In mid-1945, the government rescinded Order L-208, but the precious metal business never fully recovered. Post-war inflation once again drove up labor and materials costs, while the price of gold remained fixed. More importantly, for lode mines closed during wartime, as geologists and mining experts Howard Brooks and Len Ramp noted, "the years of idleness had resulted in such deterioration of plants and workings that prohibitively large expense would have been required for rehabilitation."[13] The passage of time only made matters worse. By 1965, the two geologists estimated that lode mining cost ten times as much as it did in 1934. Most of Oregon's post-war gold came from the Sumpter Valley Dredge, which ceased working in the 1950s. Few lode mines have operated in Oregon since World War II, and only one—the Buffalo in the Granite district—had consistent returns.

When President Nixon took the United States off the gold standard in 1971, the price of gold immediately rose, offering hope that this might spur renewed mining in Oregon. Anticipating that this event would eventually occur, Brooks and Ramp in 1965 offered a note of caution for Oregon's mining situation:

> It seems probable that the larger placer deposits of the state, particularly dredgeable areas, are, for the most part worked out or so seriously depleted that under any price structure that can reasonably be anticipated for the future, placer mining cannot be expected to resume its pre-war importance. On the other hand, there are many lode deposits in the state that have not been fully investigated. . . . Unfortunately, the important workings of nearly all of Oregon's lode gold mines are now caved or inaccessible. Consequently, the search for new ore will require costly rehabilitation or driving of new accessways into deeper or lateral portions of previously worked ore bodies. At some mines important information can be acquired by drilling from the surface or from accessible underground stations, but this is at best only a preliminary step in the development of ore bodies and must, where warranted, be followed by underground work.[14]

An added deterrent to the revival of large-scale placer or lode mining in Oregon has been the effect of federal and state environmental laws and regulations enacted in the late 1960s and since. The requirements of the National Environmental Protection Act (1969), the Endangered Species Act (1973), and the Clean Water Act (1972 and 1977), at the very least, are an added expense to mining operations. Federal agencies—including the Forest Service, Bureau of Land Management, Bureau of Reclamation, Army Corps of Engineers, and Environmental Protection Agency—oversee the implementation of these environmental laws to ensure that prospecting and mining operations follow the mandated legal requirements. In addition, state agencies and county planning commissions apply Oregon land use and clean water measures in permitting and regulating mining activities within their jurisdictions.[15]

Remarkably, even though the US Supreme Court has upheld federal and state environmental regulations affecting mining practices, other aspects of the Mining Law of 1872 have withstood all efforts to revise or repeal them. This is especially true regarding the act's hard rock mineral patent and annual work requirements, although the imposed annual $100 work assessment per claim has been changed to a simple annual $100 fee per claim. This has remained the case, even though, as a Government Accounting Office report observed in 1986:

> The patent provisions of the Mining Law of 1872 clearly run counter to other national natural resources policies and legislation. The Federal Land Policy and Management Act of 1976 (FLPMA) provides that, in general, public lands should remain under federal ownership and be managed for the benefit of all users (multiple use) as well as for future generations (sustained yield). However, mining claim holders can gain title to federal lands by patenting their claims, thereby precluding future public use of these lands.[16]

Persistent and effective lobbying by the mining industry over the last forty years has defeated the efforts to repeal the Mining Law of 1872, but environmental regulations and tribal concerns related to sacred places have changed the business conditions for precious metal mining. Also, the costs of meeting Superfund requirements for past mining environmental consequences and the new stipulations for the remediation or rehabilitation of future mining sites all factor into whether to undertake a mining project.

For over 6,000 years, humankind has been transfixed by the beauty, brightness, durability, and rarity of gold. Over time, it assumed a powerful cultural role as a universal medium of exchange. Randomly distributed throughout the earth's crust, gold made up only one millionth of 1 percent of the rocky matter comprising that crust.[17] In 1862, when prospectors discovered gold in eastern Oregon, the lure of that precious metal attracted thousands of hopeful miners to Auburn and Canyon City. First, placer and then lode mining rewarded their efforts. Between 1862 and 1910, gold and silver extraction yielded approximately $124 million (historical value), while helping usher the state into the modern industrial world. During that same time, Oregon's population grew from 52,465 to 672,765, an increase of 1,182 percent. While gold and silver mining played an important role in that growth, especially in the early years of placer mining, agriculture and, later, the timber industry formed the mainstays of Oregon's wealth. At a key stage in Oregon's early statehood, a flood of gold fired up the financial expansion of Portland and jump-started the settlement of eastern Oregon. Later, lode mining would bring another burst of outside investment to the state's economy. Unfortunately, gold mining also had negative impacts on the Native Americans use of their traditional lands and resources. In addition, the major role of the Chinese miner in the gold economy of eastern Oregon should not be forgotten. Oregon's precious metal deposits—along with its other abundant natural resources, such as timber, fish, grazing, and farmland—have contributed to the state's economic growth over the past 165 years. The thirst for gold, however, made gold stand apart from the other resources in its power to seize men's imaginations and drive their exertions.

Notes

INTRODUCTION

1 Rodman Paul and Elliott West, *Mining Frontiers of the West, 1848–1880* (Albuquerque, NM: University of New Mexico Press, rev. and exp., 2001), 149; Paul Mitchell Marks, in her survey of western gold rushes in the late nineteenth century, dismisses Oregon placer mining in one sentence: "In Oregon . . . gold strikes on the John Day River and Powder River in 1861 drew more than a thousand seekers to jockey for claims." *Precious Dust: The Saga of the Western Gold Rushes* (Lincoln, NE: University of Nebraska, 1994), 38; for a similar view downplaying the importance of Oregon mining, see also Thomas Rickard, *A History of American Mining* (New York: McGraw-Hill Books, 1932), 316.

2 Howard C. Brooks and Len Ramp, *Gold and Silver in Oregon* (Portland, OR: State of Oregon Department of Geology and Mineral Industries, 1968), 7–27, 167–169, 282; Albert Burch, "Development of Metal Mining in Oregon," *Oregon Historical Quarterly*, vol. 43 (Jun. 1941), 110; Miles F. Potter, *Oregon's Golden Years: Bonanza of the West* (Caldwell, ID: Caxton Printers, 1976), 168–169.

3 David Manuel Inflation Calculator, www.davemanuel.com/inflation, accessed April 12, 2022.

4 This and the next four paragraphs are based on the following sources: William F. Willingham, *Starting Over: Community Building on the Eastern Oregon Frontier* (Portland, OR: Oregon Historical Society, 2005), 124; William F. Willingham, *Collegiate Architecture and Landscape in the West: Willamette University, 1842–2012* (Salem, OR: Oregon State University Press, 2019), 54; *Biennial Report of the Bureau of Labor Statistics of the State of Oregon, 1905* (Salem OR: State of Oregon Printer, 1905), 58–59; Baker City Chamber of Commerce, *The Gold Fields of Eastern Oregon* (Baker City: no publisher, 1900), 50; *The Sears, Roebuck Catalogue*, 1902 ed. [reprint] (New York: Bounty Books, 1969).

CHAPTER ONE

1 Native place names throughout are sourced in Eugene S. Hunn, E. Thomas Morning Owl, Philip E. Cash, and Jennifer Carsun Engum, eds., *Caw Pasa Laakni, They Are Not Forgotten: Sahaptin Place Names Atlas of the Cayuse, Umatilla, and Walla Walla* (Pendleton and Portland, OR: Tamastslikt Cultural Institute in association with Ecotrust, 2015).

2 Verne Bright, "Blue Mountain Eldorados: Auburn, 1861," *Oregon Historical Quarterly*, vol. 62 (Sep. 1961), 213–215; Isaac Hiatt, *Thirty-One Years in Baker County: A History of the County from 1861-1893* (Baker City, OR: Abbott and Foster, 1893), 5–8; Miles F. Potter, *Oregon's Golden Years: Bonanza of the West* (Caldwell, ID: Caxton Printers, 1976), 3–8.

3 "Letters from the Upper Columbia No. 14," *Oregonian*, Jun. 27, 1861; on Pacific Northwest gold fever, see Carlos A. Schwantes, *Long Day's Journey: The Steamboat and Stagecoach Era in the Northern West* (Seattle: University of Washington Press, 1999), 108–121 and Dorothy O. Johannsen and Charles Gates, *Empire of the Columbia: A History of the Pacific Northwest*, 2nd ed. (New York, Harper and Row, 1967), 265–268: D. W. Meining, *The Great Columbia Plain: A Historical Geography, 1805–1910* (Seattle, WA: University of Washington Press, 1968), 208–219.

4 Bright, "Blue Mountain Eldorados," 215–220; Hiatt, *Thirty-One Years in Baker County*, 8–10.

5 One estimate credits southwestern Oregon placer mines with roughly $31 million in gold between 1851 and 1861; Potter, *Oregon's Golden Years*, 13–31; Howard C. Brooks and Len Ramp, *Gold and Silver in Oregon* (Portland, OR: State of Oregon Department of Geology and Mineral Industries, 1968), 8–9.

6 Bright, "Blue Mountain Eldorados," 220–223; Hiatt, *Thirty-One Years in Baker County*, 10–13.

7 Articles quoted in Bright, "Blue Mountain Eldorados," 224; Rodman Paul and Elliott West, *Mining Frontiers of the Far West, 1848–1880* (Albuquerque, NM: University of New Mexico Press, rev. and exp., 2001), 23–24, 169–171.

8 Bright, "Blue Mountain Eldorados," 225–26; Hiatt, *Thirty-One Years in Baker County*, 14–18; Virginia Duffy McLoughlin, "Cynthia Stafford and the Lost Mining Town of Auburn," *Oregon Historical Quarterly*, vol. 98 (Spring 1997), 6–17, 25–33.

9 Bright, "Blue Mountain Eldorados," 226.

10 Ibid., 227.

11 Ibid., 227.

12 Bright, "Blue Mountain Eldorados, 228; W. H. Packwood, a founding businessman of Auburn also later wrote that "at one time, in 1862 and 1863, Auburn had 40 stores and saloons;" quoted in Rossiter Raymond, *Mining Statistics West of the Rocky Mountains*, 42nd Cong., 1st sess., House Ex. Doc. 10 (Washington DC: Government Printing Office, 1872), 180; *Engineering and Mining Journal*, Feb. 28, 1871; Paula Mitchell Marks, *Precious Dust: The Saga of the Western Gold Rushes* (Lincoln, NE: University of Nebraska Press, 1994), 189–93; Paul and West, *Mining Frontiers*, 206–217; Merle W. Wells, *Gold Camps and Silver Cities: Nineteenth-Century Mining in Central and Southern Idaho* (Moscow, ID: University of Idaho Press, 2002), 3–12; see also, Kent A. Curtis, *Gambling on Ore: The Nature of Metal Mining in the United States, 1860–1910* (Boulder, CO: University Press of Colorado, 2013), 30–52.

13 As stated in the introduction, to convert historical dollar values into today's money, I have applied an inflation factor of 18.11 for the 1860s and 26.24 for the period between 1870 and 1910, derived from the David Manuel Inflation Calculator, www.davemanuel.com/inflation-calculator, accessed Nov. 5, 2024. Unless otherwise indicated, all stated dollar values in this study represent historical dollars.

14 Bright, "Blue Mountain Eldorados," 228; Hiatt, *Thirty-One Years in Baker County*, 25–31; McLoughlin, "Cynthia Stafford and the Lost Mining Town of Auburn," 13, 16; *Oregonian*, Sep. 21, 1885; Raymond, *Statistics of Mines and Mining*, House Ex. Doc. No. 10, 181; Simeon G. Reed Collection, Oregon Historical Society, "Autobiography," Jul 24, 1888, Mss. 1117, Box 7, Vol. 23, part 2, 214–217; *Engineering and Mining Journal*, Feb. 28, 1871.

15 Trimble quote in William J. Trimble, *The Mining Advance into the Inland Empire* (Madison, WI: Bulletin of the University of Wisconsin, 1914), 127; Kamm quote in E. Kimbark MacColl, *The Shaping of a City: Business and Politics in Portland, Oregon*

1885 to 1915 (Portland, OR: The Georgian Press Co., 1976), 17; Schwantes, *Long Day's Journey*, 127–137, 291–292, 301–304; Johannsen, *Empire of the Columbia*, 279–284; E. Kimbark MacColl and Harry H. Stein, *Merchants, Money, and Power* (Portland, OR: The Georgian Press, 1988), 121–130, 139; Randall V. Mills, *Sternwheelers up Columbia: A Century of Steamboating in the Oregon Country* (Lincoln, NE: University of Nebraska Press, 1947), 39–50; Arthur Throckmorton, *Oregon Argonauts: Merchant Adventurers on the Western Frontier* (Portland, OR: Oregon Historical Society Press, 1961), 247–276; Mychal Ostler, *Battle for the Columbia River: The Rise of the Oregon Steam Navigation Company* (Charleston, SC: The History Press, 2023), 1–141; E. W. Wright, ed., *Lewis and Dryden's Marine History of the Pacific Northwest* (Portland, OR: The Lewis and Dryden Printing Co., 1895), 106–130, 268; Irene Poppelton, "Oregon's First Monopoly: The OSN Co.," *Oregon Historical Quarterly*, vol. 9 (Sep. 1908), 274–304; Potter, *Oregon's Golden Years*, 31–49.

16 William F. Willingham, *Army Engineers and the Development of Oregon: A History of the Portland District, US Army Corps of Engineers* (Washington, DC: Government Printing Office, 1983), 11–13, 19–21, 24–25, 28–36, 50–51, 73–79; William F. Willingham, "Engineering the Cascades Canal and Locks, 1876–1896," *Oregon Historical Quarterly*, vol. 80 (Fall 1987), 228–257.

17 Schwantes, *Long Day's Journey*, 139–49; Potter, *Oregon's Golden Years*, 53–56.

18 Anonymous, *An Illustrated History of Baker, Grant, Malheur, and Harney County* (Spokane, WA: Western Historical Publishing Company, 1902), 382.

19 Quoted in Helen B. Rand, ed. *Whiskey Gulch, 1862–1863: Letter of G. I. Hazeltine and Wife Emeline* (Baker City, OR: The Record-Courier, 1981), 5, 17; Anonymous, *An Illustrated History*, 381–83; Potter, *Oregon's Golden Years*, 61–68.

20 John Day Mining District regulations are reprinted in Rand, *Whiskey Gulch*, 107–8; see also *The City Journal* (Canyon City), Jan. 1, 1869; for early-day Canyon City, see also Rand, *Whiskey Gulch*, 86–91; Anonymous, *An Illustrated History*, 386–387.

21 Brooks and Ramp, *Gold and Silver in Oregon*, 8–9, 45; Rossiter Raymond, *Statistics of Mines and Mining West of the Rocky Mountains*, 41st Cong., 2nd Sess., House Ex. Doc. No. 207 (Washington, DC: Government Printing Office, 1870), 224; Waldemar Lindgren, *The Gold Belt of the Blue Mountains of Oregon*, 22nd Ann. Report, Pt. 2, US Geological Survey (Washington, DC: Government Printing Office, 1901), 717.

22 Quote in Rand, *Whiskey Gulch*, 27; Hiatt, *Thirty-One Years in Baker County*, 14–41; Potter, *Oregon's Golden Years*, 61–68; Rand, *Whiskey Gulch*, 11–80, 86–96; *Blue Mountain Eagle* (May 28, 1915, Jun. 9, 1916, Oct. 14, 1921, Mar. 10, Apr. 28, May 17, Jun. 7, 1922).

23 Quotes in Bright, "Blue Mountain Eldorados," 230, 232; Hiatt, *Thirty-One Years in Baker County*, 22–23, 35–37; Anonymous, *An Illustrated History*, 150–152.

24 Harry N. M. Winter, ed., "The Powder River and John Day Mines in 1862: Diary of Winfield S. Ebey," *Pacific Northwest Quarterly*, v. 34 (Jan. 1943), 86.

25 *Blue Mountain Eagle*, Mar. 10, 1922.

26 Anonymous, *An Illustrated History*, 385; Marks, *Precious Dust*, 256.

27 This and the next paragraph are based on the following sources: Howard McKinly Corning, ed., *Dictionary of Oregon History* (Portland, OR: Binfords and Mort, 1956), 17, 43, 102, 139, 194, 250, 252, 258; Samuel N. Dicken and Emily F. Dicken, *The Making of Oregon: A Study in Historical Geography* (Portland, OR: Oregon Historical Society Press, 1979), 84–85; Lewis A. McArthur, *Oregon Geographic Names* (Portland, OR: Oregon Historical Society Press, 2003), 157; Keith May and Christina Rae, *Pendleton: A Short History of a Real Western Town* (Pendleton, OR; Drigh Sighed Publications, 2005), 7–8; Gordon MacNab, *A Century of News and People in the East Oregonian, 1875–1975* (Pendleton, OR: East Oregonian Publishing Co., 1975), 16–25.

28 Quote in W. Turrentine Jackson, *Wagons Roads West: A Study of Federal Road Surveys and Construction in the Trans-Mississippi West* (Lincoln, NE: University of Nebraska Press, 1964), 86, see also 84–88; Anonymous, *An Illustrated History*, 379–380; US Army

Topographical Engineer, Capt. Thomas Cram produced a comprehensive report on the topography of the Oregon territory in March 1859. His study focused on transportation and military conditions in the region, noting that the existing road (Oregon Trail) across the Blue Mountains "is hard upon animals and wagons" and the Grand Rond[e] "is the best valley in the whole country . . . [and] the favorite summer resort of several tribes of Indians." *Topographical Memoir of the Department of the Pacific*, 35th Cong. 2nd Sess., 1859, House Ex. Doc. No. 114 (Washington, DC: Government Printing Office, 1859), 59; Priscilla Knuth, "Images of the Deschutes Country," in Thomas Vaughan, ed., *High and Mighty: Select Sketches about the Deschutes Country* (Portland, OR: Oregon Historical Society Press, 1981), 162–165.

29 Anonymous, *An Illustrated History*, 388, 391.

30 Anonymous, *An Illustrated History*, 388, 395, 406–08; Jerry A. O'Callaghan, *The Disposition of the Public Domain in Oregon* (New York: Arno Press, 1979), 49–50, 54–55; W. Turrentine Jackson, "Federal Road Building Grants for Early Oregon," *Oregon Historical Quarterly*, vol. 50 (Mar. 1949), 26–27, 29.

31 Jennifer Karson, ed., *Wiyáxayxt/Wiyáakaa'awn = As Days Go By: Our History, Our Land, Our People—The Cayuse, Umatilla, and Walla Walla* (Portland, OR: Oregon Historical Society Press, 2006), 65–89; Omar C. Stewart, "The Northern Paiute Bands," in *Anthropology Records* (Berkeley, CA: University of California Press, 1939), Map 1, 127–148; Hiatt, *Thirty-One Years in Baker County*, 61–73; Anonymous, *An Illustrated History*, 154–156, 389–393; Mike Higgins and Les Tipton, *Ditch Walkers and Water Wars: The Life and Times of the Eldorado Ditch in the Gold Fields of Eastern Oregon* (Caldwell, ID: Blue Pine Publishing Company, 2013), 14; *Blue Mountain Eagle*, Mar. 17, 1922.

32 First quote in Priscilla Knuth, ed. "Cavalry in the Indian Country, 1864," *Oregon Historical Quarterly*, vol. 65 (Mar. 1964), 117, for Drake's journal of the campaign, see 5–118; Anonymous, *An Illustrated History*, 390, 393; James Robbins Jewell, ed., *On Duty in the Pacific Northwest During the Civil War: Correspondence and Reminiscences of the First Oregon Cavalry Regiment* (Knoxville, TN: University of Tennessee Press, 2018), 83–150, 165–197; Peter Cozzens, ed., *Eyewitnesses to the Indian Wars, 1865–1890: The Wars for the Pacific Northwest* (Mechanicsburg, PA: Stackpole Books, 2002), 10–92; David Wilson Jr., *Northern Paiutes of the Malheur* (Lincoln, NE: University of Nebraska Press, 2022), 44–46. For the Malheur Reservation, see, Jeff Zucker, Kay Hummel, and Bob Hogfoss, *Oregon Indians: Culture, History and Current Affairs* (Portland, OR: Oregon Historical Society Press, 1983), 103–5; For the Bannock War, see William F. Willingham *Starting Over: Community Building on the Eastern Oregon Frontier* (Portland, OR: Oregon Historical Society Press, 2006), 30–37 and Anonymous, *An Illustrated History*, 743–754, 776–780; Cozzens, ed., *Eyewitnesses to the Indian Wars*, 604–674; *Blue Mountain Eagle*, Mar. 17, 1922, Dec. 4, 1925, Jun. 17, 1932; second quote in James Robbins Jewell, *Agents of Empire: The First Oregon Cavalry and the Opening of the Interior Pacific Northwest during the Civil War* (Lincoln, NE: University of Nebraska, 2023), 271.

33 Hiatt, *Thirty-One Years in Baker County*, 52–61, 78–79; Anonymous, *An Illustrated History*, 153, 155, 159–162, 205–208, 393–397, 431–434; Potter, *Oregon's Golden Years*, 95–97.

34 Meining, *The Great Columbia Plain*, 294–320, 411–413, 420, 500; Willingham, *Starting Over*, 13–62; William Parsons, *An Illustrated History of Umatilla County* (W. H Lever, 1902), 160–162.

35 *Oregon House and Senate Journals*, 4th Session (Salem, OR, 1866), 50; *Biennial Report of Secretary of State* (Salem, OR: 1876), 13.

36 On the rise of the livestock industry, see Willingham, *Starting Over*, 41–54. The southern portions would become Malheur (1887) and Harney (1889) counties; the state census data can be found at *Oregon House and Senate Journals*, 51–52; *Secretary of State Report*, 14–15.

37 Schwantes, *Long Day's Journey*, 139–42, 303–304; Willingham, "Engineering the Cascades Canal," 230–233; D. W. Meinig, *The Great Columbia Plain*, 223–234, 249–257; *Compendium of the Ninth United States Census, 1870* (Washington, DC: Government

Printing Office, 1872), 772–773; *Compendium of the Tenth United States Census, 1880* (Washington, DC: Government Printing Office, 1885), 806.

38 Frederic Trautmann, ed., *Oregon East, Oregon West: Travels and Memoirs by Theodor Kirchhoff, 1863–1872* (Portland, OR: Oregon Historical Society Press, 1987), 157–158.

39 John M. Murphy, compiler. *Oregon Business Directory and State Gazetteer, 1873* (Portland, OR: S. J. McCormick, 1873), 65.

40 Brooks and Ramp, *Gold and Silver in Oregon*, 7–9; Potter, *Oregon's Golden Years*, 41, 49, 59–60, 168–169; Trimble, *The Mining Advance into the Inland Empire*, 73–74, 102, 118; Leslie M. Scott, "The Pioneer Stimulus of Gold," *Oregon Historical Quarterly*, vol. 18 (Sept. 1917), 147–66; Walter R. Crane, *Gold and Silver: An Economic History of Mining in the United States* (New York: John Wiley and Sons, 1908), 625, 628; J. Arthur Phillips, *Mining and Metallurgy of Gold and Silver* (London: E. and F. N. Spon, 1866), 76; Lindgren, *The Gold Belt*, 569–573; Raymond, *Statistics of Mines and Mining*, House Ex. Doc. 207, 224.

41 Hiatt, *Thirty-One Years in Baker County*, 40–46, 60–63, 78–79; Meinig, *The Great Columbia Plain*, 209–256; Willingham, *Starting Over*, 13–51; *Compendium of the Tenth United States Census, 1880*, 48; Oregon State Census, 1865, 50.

42 This and the following paragraph are based on MacColl and Stein, *Merchants, Money, and Power*, 121–61, 181–184; Throckmorton, *Oregon Argonauts*, 247–294; Trimble, *The Mining Advance into the Inland Empire*, 114–115.

43 Quoted in Carl Abbott, *Portland in Three Centuries: The Place and the People*, 2nd ed. (Corvallis, OR: Oregon State University Press, 2022), 42.

44 Edward G. Jones, *The Oregonian's Handbook of the Pacific Northwest* (Portland, OR: Oregonian Publishing Co., 1894), 111.

45 "Banks of Portland, The Financial Stronghold of the Northwest," *Oregonian*, Jul. 3, 1897.

46 J. H. Fisk, *The Mineral Resources of Oregon* (Portland, OR: Lewis and Clark Centennial Exposition Commission, 1904), 7–8.

47 Wells, *Gold Camps and Silver Cities*, 36–38, 41; Richard H. Peterson, *The Bonanza Kings*, (Norman, OK: University of Oklahoma Press, 1991), 26, 56–61, 92, 104, 126, 129-131, 141; Reed, "Autobiography," and Boxes 2-7, passim. The Reed Collection contains monthly reports and letters to and from the mine managers and Reed over the course of a decade. All are arranged chronologically and interspersed with all his other correspondence. The collection is voluminous but well-indexed.

48 Quote in MacColl and Stein, *Merchants, Money, and Power*, 153.

49 MacColl and Stein, *Merchants, Money, and Power*, 143–146; to place this asset evaluation in perspective, a typical worker averaged an income of about $600 a year.

50 Quote in Throckmorton, *Oregon Argonauts*, 276.

51 Willingham, *Army Engineers and the Development of Oregon*, 10–26; MacColl and Stein, *Merchants, Money, and Power*, 181–184; Throckmorton, *Oregon Argonauts*, 295–315; E. W. Wright, ed., *Marine History*, 248–249; *Oregonian*, Jan. 1, 1898; for population figures, see US Census 1860, and Bureau of the Census, *Thirteenth Census of the United States, Supplement for Oregon* (Washington DC: Government Printing Office, 1913), 602.

52 This and the next paragraph are based on the following: Hiatt, *Thirty-One Years in Baker County*, 20–21, 32, 86; McLoughlin, "Lost Mining Town of Auburn," 34–42; Rand, *Whiskey Gulch*, 86–94; Anonymous, *An Illustrated History*, 180–181, 188–200, 383–384, 394, 416, 426–434; Trimble, *The Mining Advance into the Inland Empire*, 146–154, 160, 168-169, 171–176, 246–247; James Waucop Tabor, *Granite and Gold* (Baker, OR: Record-Courier Printers, 1988), 62–63; Paul and West, *Mining Frontiers*, 138–419, 211–225; Marks, *Precious Dust*, 247–277; Curtis, *Gambling on Ore*, 48–56; Johannsen and Gates, *Empire of the Columbia*, 268–273; Christian Spreen, "A History of Placer Gold Mining in Oregon, 1850–1870 (MA Thesis, University of Oregon, 1939), 70; Carlos Schwantes, *The Pacific Northwest: An Interpretive History* (University of Nebraska Press, 1996), 133–140; Richard White, *It's Your Misfortune and None of My Own": A*

New History of the American West (Norman, OK: University of Oklahoma Press, 1991), 171–172, 192; Clyde A. Miner, Carol A. O'Connor, and Martha A. Sandweiss, eds, *The Oxford History of the American West* (New York: Oxford University Press, 1994), 183–184, 188, 204–205.

53 Benjamin Mountford, "The Pacific Gold Rushes and the Struggle for Order," in Benjamin Mountford and Stephen Tuffnell, eds., *A Global History of Gold Rushes* (Oakland, CA: University of California Press, 2018), 88–108.

CHAPTER TWO

1 The following discussion of Blue Mountain geology is based on Brooks and Ramp, *Gold and Silver in Oregon*, 41–53; Howard Brooks, *A Pictorial History of Gold Mining in the Blue Mountains of Eastern Oregon* (Baker City, OR: Baker County Historical Society, 2007), 23–28; Elizabeth Orr and William Orr, *Geology of Oregon*, 6th ed. (Corvallis, OR: Oregon State University Press, 2012), 20–47; US Department of the Interior, Geological Survey, *The Geologic Setting of the John Day Country* (Washington, DC: Government Printing Office, 1970), 2–23.

2 *Mining and Scientific Press* (hereafter *M&SP*), Dec. 3, 1904.

3 Raymond, *Statistics of Mines and Mining West of the Rocky Mountains*, House Ex. Doc. No. 207, 475.

4 *West Shore*, Apr. 1887, 310; see also, Andrew C. Isenberg, *Mining California: An Ecological History* (New York: Hill and Wang, 2005), 23–51; Otis E. Young, Jr., *Western Mining* (Norman, OK: University of Oklahoma Press, 1970), 108-136.

5 Raymond, *Statistics of Mines and Mining West of the Rocky Mountains*, House Ex. Doc. No. 207, 475–476, 693-694.

6 Quote in Trimble, *The Mining Advance Into the Inland Empire*, 95.

7 "The Development of the Metal Mining Industry in the Western States," *M&SP*, Dec. 1, 1906.

8 Elliott West, *Continental Reckoning: The American West in the Age of Expansion* (Lincoln, NE: University of Nebraska Press, 2023), 387.

9 There were three different dredges built to operate in the Sumpter Valley over the period between 1913 and 1954. Brooks Hawley, *Gold Dredging in Sumpter Valley* (Baker City, OR: Baker Printing and Lithography, 1977), quotes on 2, see also, 11–12, 19–28; Brooks, *A Pictorial History of Gold Mining*, 42–44.

10 The description of lode mining methods in the next few paragraphs is based on Brooks, *Pictorial History of Gold Mining*, 29–45; Trimble, *The Mining Advance Into the Inland Empire*, 90–100; John Cumming, *Mining Explained: A Layperson's Guide* (Toronto, ON: The Northern Miner, 2012), 39–64; "The Free-Milling Process," *M&SP*, Dec. 31, 1904; "Principles Underlying the Amalgamation of Gold in the Stamp-Mill," *M&SP*, Jul. 7, 1906; "The Development of the Metal Mining Industry in the Western States," *M&SP*, Dec. 1, 1906; J. S. Phillips, *The Explorers' Miners' and Metallurgists' Companion* (San Francisco, CA: Chas, W. Gordon, Book and Job Printer, 1878), 514–555; W. H. Tinney, *Gold Mining Machinery: Its Selection, Arrangement & Installation* (New York: D. Van Nostrand, 1906), 179–212; Crane, *Gold and Silver*, 343–552; Young, *Western Mining*, 136–140, 279–285; Bureau of the Census, *Mines and Quarries, 1902* (Washington, DC: Government Printing Office, 1905), 593–602.

11 West, *Continental Reckoning*, 386.

12 Mark Wyman, *Hard Rock Epic: Western Miners and the Industrial Revolution, 1860–1910* (Berkeley, CA: University of California Press, 1979), 81–87, 89, 102–114; Eric C. Nystrom, *Seeing Underground: Maps, Models, and Mining Engineering in America* (Reno, NV: University of Nevada Press, 2014), 7–8; *Blue Mountain Eagle* (Mar. 15, Jul. 5, 1901).

13 Thomas T. Cook, *The Cornucopia: Oregon's Richest Gold Mine* (no publisher, 2016), 2.

14 Trimble, *The Mining Advance Into the Inland Empire*, 95–100.

15 *Sumpter Miner* (Jan. 13, Mar. 30, 1904). The newspaper did cover union social events: See, for example, Jan. 31, 1900; Oct. 8, 1902; and Apr. 29, 1903. Miners' union activities in the West are covered in Wyman, *Hard Rock Epic*, 84–145; Paul and West, *Mining Frontiers*, 270–81. In eastern Oregon, one mine laborer accidentally fell to his death inside the Connor Creek Mine, as reported in a letter from the mine superintendent, J. Myrick, to the mine owner, Simeon Reed, May 30, 1887, Simeon G. Reed Collection, Oregon Historical Society, Mss. 1117, Box 6, Vol. 18. The *Grant County News*, Nov. 10, 1904, reported that a mine worker was killed in an underground accident at the Badger Mine.

16 This and the following two paragraphs are based on Stephen Tuffnell, "Engineers and the Mechanics of Global Connectivity," in Mountford and Tuffnell, eds., *A Global History of Gold Rushes*, 229–251; Nystrom, *Seeing Underground*, 8–10, 46–49, 57–59; Clark C. Spence, *Mining Engineers and the American West: The Lace-Boot Brigade, 1849–1933* (New Haven, CT: Yale University Press, 1970), 18–53, passim; Curtis, *Gambling on Ore*, 83–97.

17 Tuffnell, "Engineers and Mechanics," 232.

18 Tuffnell, "Engineers and Mechanics," 237.

19 Quotes in Curtis, *Gambling in Ore*, 89, 90.

20 This and the following two paragraphs are based on Gordon Morris Bakken, *The Mining Law of 1872: Past, Politics, and Prospects* (Albuquerque, AZ: University of New Mexico Press, 2008), 1–31; Curtis, *Gambling on Ore*, 78–83, 97–101. The editor of the M&SP (Aug. 11, 1877 issue) pointed out that alien Chinese could not make valid mining claims under the terms of the Mining Law of 1872.

21 Bakken, *The Mining Law of 1872*, 26.

22 *M&SP*, May 21, 1864; see also, Young, *Western Mining*, 18–32.

23 *M&SP*, Apr.1, 1882, see also Jun. 14, 1884.

24 Curtis, *Gambling on Ore*, 104.

CHAPTER THREE

1 This and the following paragraph are based on Samuel Dicken and Emily Dicken, *Oregon Divided: A Regional Geography* (Portland, OR: Oregon Historical Society Press, 1982), 119–122.

2 Eugene McCornack, Cadastral Survey Notes, T.8S., R.351/2E. (Portland, OR: General Land Office [Bureau of Land Management Archives], 1881).

3 H. Gradon, Cadastral Survey notes, T.10S., R.32E. (Portland, OR: General Land Office [Bureau of Land Management Archives], 1881).

4 The discussion of Native American use of the Blue Mountain region is based on the following sources: Karson, ed., *As the Days go By*, 5, 7, 23–57, 65–86; Terry Ozburn, et al. *Archaeological Data Recovery at the Private Trust Site (35GR1689), the East Big Boulder Creek Site (35GR1730)*, and *Treatment of Two Segments of the Sumpter Valley Railway Middle Fork John Day River Highway Project, Grant County, Oregon*, Report No. 130 Portland, OR: Archaeological Investigations Northwest, 1997), 1–34; Norm Steggell, "An East Central Oregon Forested Highland Occupation from c. 10,000 BP to Historic Times," paper presented at the Great Basin Anthropological Conference, Reno, Nevada, Oct. 2, 1982; Kathryn Anne Toepel, William F. Willingham, and Rick Minor, *Cultural Resources Overview of BLM Lands in North-Central Oregon* (Eugene, OR: Dept. of Anthropology, University of Oregon, 1979), 26–126; C. Melvin Aikens, *Archaeology of Oregon* (Portland, OR: US Dept. of the Interior, Bureau of Land Management, Oregon State Office, 1993), 13–134; Jeff Zucker, Kay Hummel, and Bob Hogfoss, *Oregon Indians: Culture, History, and Current Affairs* (Portland, OR: Western Imprints, the Press of the Oregon Historical Society, 1983), 9, 11–12, 29–31, 38–9, 46, 48–53, 57; Wilson, *Northern Paiutes*, 5–30.

5 Anonymous, *An Illustrated History*, 392 (quote), 389–93, 520, 633–34. See also, *Grant County News*,Apr 1. 1886, Aug 6, 1885, Feb. 16, 1888; *Blue Mountain Eagle*, Mar. 17, Apr. 7, 1922, May 31, 1962, Oct. 9, 1975; *Oregonian*, Sep. 17, 1866.

6 Anonymous, *An Illustrated History*, 393.

7 Lewis A. McArthur and Lewis L. McArthur, *Oregon Geographic Names*, 7th ed. (Portland, OR: Oregon Historical Society Press, 2003), 148; Jeanne Secord, compiler, Yesterday in Grant County (John Day, OR: n. p., 1973), 47–51, 59–60; *Oregonian*, Sep. 11, 17, Dec. 3, 1866, Jan. 14, Apr. 1, May 21, Jul. 20, Aug. 31, Sep. 19, Oct. 1, Nov. 1, 1867, Sep. 18, Dec. 12, 1868. Camp Logan was garrisoned temporarily during 1865, *Daily Mountaineer* (The Dalles), Nov. 3, 1865; Wilson, *Northern Paiutes*, 23–86.

8 For more on the Bannock War in Grant County, see Willingham, *Starting Over*, 30–37; Anonymous, *An Illustrated History*, 743–754, 776–780; Cozzens, ed., *Eyewitness to the Indian Wars*, 604–674; Wilson, *Northern Paiutes*, 87–225.

9 James Waucop Tabor, *Granite and Gold* (Baker, OR: Record-Courier Printers, 1988), 6–9. A somewhat different account of the discovery of gold on Granite Creek appeared in the *M&SP*, Aug. 25, 1877. In part, it read: On July 3, 1862, the mule of miner, Jack Long, got "mired in Granite Creek, whereupon he returned for a pick and shovel, declaring his intention to sink a hole under that mule for luck; the remaining members of the party jokingly encouraged him. Jack was as good as his word; after extricating the mule, he continued sinking to the depth of about four feet, where he washed out [gold worth] 25 cents to the pan. This excited everybody, claims were staked off immediately, and a town laid out, which was very appropriately named Independence, it having been located on Independence Day."

10 Tabor, *Granite and Gold*, 10, 15; Howard Brooks, *Pictorial History of Gold Mining*, 92; Raymond, *Statistics of Mines and Mining*, House Ex. Doc. No. 207, 224.

11 *Oregonian*, Oct. 9, 1862.

12 *Oregonian*, Jun. 10, 1863; see also *Oregonian*, Jul. 23, 1863.

13 *Oregonian*, Sept. 14, 1865.

14 *M&SP*, Jul. 21, 1866.

15 *M&SP*, Mar. 28, 1868; Grant County Mining Claim Records, Book B, 1866–1881 (Canyon City, OR: Grant County Court House).

16 Murphy, *Oregon Business Directory and State Gazetteer, 1873*, 64.

17 *Bedrock Democrat* (Baker City), Jun. 9, 1875; see also Jul. 15, 1874.

18 Tabor, *Granite and Gold*, 59-61; *East Oregonian*, Feb. 17, Mar. 31, Jul. 23, 1877; *Bedrock Democrat* (Baker City), Jul. 15, 1874.

19 US Bureau of the Census, *Ninth Census, The Statistics of the Wealth and Industry of the United States* (Washington, DC: Government Printing Office, 1872), 759-785.

20 *East Oregonian*, Mar. 31, 1877.

21 Anonymous, *An Illustrated History*, 168.

22 *Oregonian*, Sept. 21, 1885; Lindgren, *The Gold Belt*, 686.

23 *East Oregonian*, Mar. 31, 1877.

24 *M&SP*, Jul. 10, 1875.

25 *M&SP*, Aug. 11, 1877.

26 *Grant County News*, Oct. 11, 1879; Chinese mine workers mentioned in the *Oregonian*, Aug. 6, 1879.

27 Hendryx File (microfilm), Oregon Historical Society.

28 U. S. Bureau of the Census, "Products of Industry Schedules, Oregon, Grant County, 1880" unpublished manuscript, Oregon State Library, Salem, Oregon.

29 Quote in *Grant County News*, May 6, 1882. The development of the Monumental Mine received heavy newspaper coverage, see *East Oregonian* Mar. 31, Apr. 14, 21, Jul. 23, 1877; Aug. 2, 9, 23, Oct. 25, Nov. 8, 15, 22, Dec. 6, 13, 1879; Mar. 13, Jul. 10, Oct. 2,

1880; *Grant County News*, Sept. 27, Oct. 11, 1879, Jan. 3, Feb. 14, 21, Mar. 13, 1880; *Oregonian*, Apr. 5,12, 10, 1877; Aug. 12, Nov. 20, 1878; May 1, Jun.4, 12, Aug. 6, 16, Nov. 25, 1879; May 11, 18, Jul. 9, Aug. 18, Dec. 25, 1880; Jul. 29, 1882; Sep. 10, 1888; Jan. 19, Dec. 16, 1893; Jan 8, 1894; Apr. 6, 14, 18, 1898; Jul. 22, 1922; *M&SP*, Sep. 2, 1876, Feb. 17, Mar. 3 and 24, Aug. 11, 25, Oct. 20, 1877, Feb. 9, 16, 23, Jul. 6, Nov. 16, Dec. 21, 1878, Aug. 30, 1879, Sep. 22, Oct. 6, 1888, Feb. 2, 9, Aug. 17, 1889, Jul 19, 1890, Jul 19, Dec. 26, 1891, Feb. 6, Mar. 19, Oct. 15, 1892, Oct. 21, 1893; *Blue Mountain American*, Mar. 7, 1903. See also, Oregon Department of Geology and Mineral Industries, *Oregon Metal Mines Handbook*, Bulletin 14-B (Salem, OR: State of Oregon, 1941), 56–57; Brooks, *A Pictorial History of Gold Mining*, 101–102.

30 *Blue Mountain Eagle*, Jun. 23, 1905 (reprint of early county records).

31 Grant County, Record of Mining Locations, Book A (Canyon City, OR: Grant County Courthouse).

32 *Oregonian*, Sep. 14, 1865.

33 *M&SP*, Jun. 9, 1866, see also Jul. 21, 1866.

34 *M&SP*, Apr. 20, 1867.

35 Quoted in *M&SP*, Jul. 6, 1867, see also Aug. 3 and Nov. 16, 1867.

36 Anonymous, *An Illustrated History*, 394.

37 *M&SP*, Mar. 28, 1868, see also Apr. 11, Oct. 3, 1868.

38 Raymond, *Statistics of Mines and Mining*, House Ex. Doc. No. 207, 222–223.

39 *M&SP*, Jan. 7, 1871.

40 U. S. Bureau of the Census, "Products of Industry Schedules, Oregon, Grant County, 1870" unpublished manuscript, Oregon State Library, Salem, Oregon.

41 Murphy, *Oregon Business Directory*, 64.

42 Raymond, *Statistics of Mines and Mining*, Ex., Doc. 207, 222-223; Rossiter Raymond, *Statistics of Mines and Mining in the States and Territories West of the Rocky Mountains*, House of Representatives, 43rd Cong, 1st Sess., House Ex. Doc. No. 141 (Washington DC: Government Printing Office, 1874), 252; *Oregonian*, Aug. 24, 1874.

43 *M&SP*, Aug. 1, 1874, see also Sep. 19, Oct. 3, 1874.

44 *Oregonian*, Apr. 21, 1875, also quoted in *M&SP*, May 8, 1875, see also Aug. 7, 1875. The *Oregonian*, gave several notices in its pages to the Cabell Mine in the mid-1870s, especially noting that the mine's eight-stamp mill machinery was of inferior capacity and therefore limiting production: Jul. 27, Aug. 10, 31, Sep. 1874, Aug. 30. 1875.

45 Burch, "Development of Metal Mining in Oregon," *Oregon Historical Quarterly*, vol. 43 (June 1942), 110–112. Simeon G. Reed Collection, Oregon Historical Society, Mss. 1117, Boxes 1-7, *passim*; Anonymous, *Illustrated History*, 755, 763; H. M. Parks and A. M. Swartley, Handbook of the Mining Industry of Oregon, in *Mineral Resources of Oregon*, Vol. 2, No. 4 (Corvallis, OR: Oregon Bureau of Mines and Geology, 1916), 68–69; Brooks, *Pictorial History*, 174–175.

46 Lindgren, *Gold Belt*, 722-723; Raymond, *Statistics of Mines and Mining*, House Ex. Doc. No. 207, 230-232, House Ex. Doc. No. 10, 185, House Ex. Doc. 141, 252-254, House Ex. Doc. No. 211, 257, House Ex. Doc. No. 210, 212-213, House Ex. Doc. No. 177, 321; *Engineering and Mining Journal*, Feb. 28, 1871; Parks and Swartley, *Mineral Resources of Oregon*, 229-30; Brooks, *A Pictorial History of Gold Mining*, 133-135; *Oregon Metal Mines Handbook*, Bulletin 14-A, 108-109; Kim and Tim Lethlean, personal communication, Jun. 6, 2024. See also *M&SP*, Jul. 2, 1870 and Jul. 22, 1876.

47 Quoted in *M&SP*, Sep. 10, 1881; see also Jan. 28, 1882.

48 Quoted in *M&SP*, Apr. 15, 1882

49 Quoted in *M&SP*, Apr. 29, 1882, see also Sep. 23, 1882; U. S. Census Bureau, *Compendium of the Tenth Census, Part 2, Mining* (Washington, DC: Government Printing Office, 1883), 1230-1235.

50 This newspaper article was based on Horatio Burchard, *Report of the Director of the Mint upon the Statistics of the Production of Precious Metals in the United States, for 1882* (Washington, DC: U. S. Government Printing Office, 1883), 186-190; Anonymous, *An Illustrated History*, 496-497.

51 Lindgren, *Gold Belt*, 707; see also, *Grant County News*, May 7, 1885.

52 For other examples, see *M&SP*, Apr. 25, May 30, Jun. 27, Dec. 19, 1885, Jan. 2, 23, May 22, 1886, Jan.29, Mar. 12, May 28, Aug. 20, 27, Oct. 29, Nov. 5, Dec. 10, 1887, Feb. 4, Jun 23, 1888, Mar. 16, May 25, Nov. 9, 30, Dec. 28, 1889; *The Pacific Miner*, Jul 1, 1902.

53 Quote in *M&SP*, Jun 23, 1888, see also Mar. 16, Sep. 7, Nov. 9, 1889.

54 For other examples of lode mining in the Granite District, see *M&SP*, Jul. 9, Dec. 10, 1887, Jan. 14, Feb. 4, 11, Mar. 3, 11, Apr. 7, Jul. 28, Aug. 4, 18, Sep. 1, Oct. 13, 20, Dec. 22, 1888, Feb. 9, Mar. 16, Apr. 6, May 25, Jun. 1, Sep. 7, Nov. 9, Dec. 28, 1889; *East Oregonian*, Jul. 19, 1888; Anonymous, *An Illustrated History*, 168, 172; Lewis A. McArthur, *Oregon Geographic Names*, 7th ed. (Portland, OR: Oregon Historical Society Press, 2003), 926.

55 *East Oregonian*, Sep. 1, 1888; see also, Jul. 19, 1888.

56 Marian V. Sears, "Jonathan Bourne, Jr., Capital Market and the Portland Stock Exchange . . . 1887," *Oregon Historical Quarterly*, vol. 69 (Sep. 1968), 197-222.

57 Quote in MacColl, *The Shaping of a City*, 204; see also *M&SP*, Oct. 18, 1902, Jan. 19, 1907.

58 Sears, "Jonathan Bourne," 197–222; MacColl, *The Shaping of a City*, 63–64, 78, 204; MacColl and Stein, *Merchants, Money, and Power*, 245–46, 261.

59 Anonymous, *Illustrated History*, 225–375, 453–574.

60 Anonymous, *Illustrated History*, 485, 496–497; Willingham, *Starting Over*, 26–27, 114, 172, 178; *Sumpter Miner*, May 22, 1901, *Grant County News*, Dec. 18, 1880, Jan. 24, Jun. 20, 1889, May 31, 1906; *Blue Mountain Eagle*, May 20, 1910. Tabor, *Granite and Gold*, 4, 6, 10, 28–31, 34.

61 *East Oregonian*, Apr. 29, 1887; for another example of combining ranch and mine work, see *Long Creek Eagle*, Dec. 2, 1892.

62 Anonymous, *Illustrated History*, 171, 405.

63 The following demographic analysis of Granite and Elk Creek (Susanville) is based on US Bureau of the Census, Ninth Census Population Schedules (Microfilm Publication Number M593), Roll 1286; US Bureau of the Census, Tenth Census Population Schedules (Microfilm Publication Number T9), Roll 1081.

64 Wyman, *Hard Rock Epic*, 41–42; Marks, *Precious Dust*, 28, 84, 128, 152, 191, 280–281; Paul and West, *Mining Frontiers*, 231, 234–236; Ronald M. James, *The Roar and the Silence: A History of Virginia City and the Comstock Lode* (Reno, NV: University of Nevada Press, 1998), 91–100, 137–139, 143–166, 238–239, 244–247.

CHAPTER FOUR

1 Erik Eklund, "Creating a Global Industry: Geology, Capital, and Company Formation on the Goldfields of the Industrial Age," in Mountford and Tuffnell, *A Global History of Gold Rushes*, 184–205.

2 *West Shore*, Jan. 11, 1890; see also, *Engineering and Mining Journal*, Jan. 3, 1891; *M&SP*, Feb. 1, 1890; Anonymous, *An Illustrated History*, 172.

3 *Bedrock Democrat*, Nov. 28, 1891; see also, *Bedrock Democrat*, Apr. 4, 1891; *Engineering and Mining Journal*, Jan. 18, Jul. 5, Sep. 6, 1890, May 16, Sep. 6, 26, 1891, Jun. 4, Jul. 16, Oct. 8, Dec. 10, 1892.

4 See also, *M&SP*, May 16, 1891, Sep. 10, Oct. 5, 1892.

5 Quoted in *M&SP*, Jul. 16, 1892; see also, *M&SP*, Jun. 4, Sep. 24, Oct. 8, 15, 29, 1892.

6 *M&SP*, Feb. 25, 1893; see also, *M&SP*, Aug. 5, 1893, and *East Oregonian*, Jan. 1, 1892.

7 Quoted in *M&SP*, Oct. 20, 1894; see also, *M&SP*, June 16, Aug. 11, 1894.

8 *Oregonian*, Jan. 1, 1898. Eastern Oregon mines received extensive coverage in the national mining press from 1895 to 1899. I found the following news accounts and articles especially useful in covering the period: *Engineering and Mining Journal*, Jan. 5, 12, 19, Feb. 28, Mar. 9, May 11, Jun. 8, 1895; Feb. 15, Apr. 18, May 30, Jun. 13, 1896; Feb. 20, Apr. 24, Jun. 19, Oct. 30, 1897; Jul. 2, Aug. 20, Sep. 17, Oct. 8, Nov. 19, Dec. 10, 1898; Jan. 7, May 20, 1899. See also, *M&SP*, Apr. 20, May 11, Jun. 1, 22, Aug. 3, 17, 1895; Feb. 22, Mar. 21, Apr. 4, May 11, Aug. 15, 22, 29, Sep. 19, Oct. 17, Nov. 7, 21, 1896; Jan. 23, Feb. 13, 27, Mar. 20, Apr. 13, 17, Aug. 14, 28, Sep. 11, Oct. 9, 23, Nov. 20, 1897; Feb. 19, Mar. 12, 18, Jul. 2, 23, Aug. 20, Sep 10, 24, Nov. 5, 12, Dec. 3, 1898; Jan. 21, Feb. 1, 18, Mar. 11, Apr. 1, Jul 22, Aug. 19, Sep. 16, 30, Oct. 7, 14, 28, Nov. 18, 25, 1899.

9 Oregonian, Jan. 1, 1897.

10 *M&SP*, Aug. 28, 1897.

11 *M&SP* Aug. 28, 1897, Jul 2, Aug. 20, 1898, Feb. 18, Aug. 19, 1899.

12 Peter. Donan, *Gold Fields of Baker County Eastern Oregon* (Portland, OR: Oregon Railroad and Navigation Co., 1897), 10.

13 *Oregonian*, Jun. 10. 1898.

14 Clark Spence, "British Investment and Oregon Mining, 1860-1900, in *Oregon Historical Quarterly*, vol. 58 (Mar. 1957), 101–112.

15 William G. Robbins, *Colony and Empire: The Capitalist Transformation of the American West* (Lawrence, KS: University Press of Kansas, 1994), 86–87.

16 *M&SP*, Aug. 19, 1899.

17 *M&SP*, Jan. 5, 1895.

18 See also, *M&SP*, Feb. 1, 1890, Nov. 18, 25, 1899.

19 *Morning Democrat, Souvenir Edition*, May 20, 1898, 2.

20 *Souvenir Edition*, 3.

21 *Souvenir Edition*, 3; Tabor, *Granite and Gold*, 10–13.

22 The discussion of the Susanville District is based on the following: Lindgren, *Gold Belt*, 705–708; Brooks and Ramp, *Gold and Silver in Oregon*, 45, 106, 108–109, 120–123, 77; *Oregon Metal Mines Handbook*, Bulletin No. 14-B (Portland, OR: Department of Geology and Mineral Industries, 1941), 129–133, 147–148; Brooks, *Pictorial History of Mining*, 72–73; *Oregonian*, Nov. 20, 1896; *Grant County News*, May 24, 31, 1888, Dec. 12, 1889, Apr. 10, 1890, Dec. 3, 1891, Jul. 2, 1898; *Long Creek Eagle*, Jul. 3, Aug. 28, Oct. 9, 23, Dec. 3, 25, 1891, Apr. 22, 29, Jun. 24, Sep. 16, Dec. 9, 1892, Jan. 20, Feb. 3, May 5, 1893, Jan. 21, 1899; Oregon Department of Geology and Mineral Industries, Oregon Historical Mining Information, Susanville General Report.

23 *Oregonian*, Mar. 18, 1894; see also, *Grant County News*, May 14, 31, 1888, Dec. 12, 1889, Apr. 10, 17, May 15, Jul. 31, Aug. 7, 1890, Apr. 9, Aug. 20, 1891, Jun. 24, Nov. 25, 1892; *Long Creek Eagle*, Dec. 3, 25, 1891, Jun. 24, 1892, Apr. 14, 1893.

24 *Grant County News*, Apr. 17, 1890. In the same article, the editor also referred to Susanville as "an ancient city which flourished twenty years with its hundreds of busy inhabitants."

25 Quote in Potter, *Oregon's Golden Years*, 77; *Oregonian*, Jul. 18, 1897.

26 *Oregonian*, Nov. 3, 1897.

27 *Oregonian*, Apr. 14, 1898.

28 Quote in the *Oregonian*, Jul. 18, 1898.

29 *Oregonian*, Mar. 18, 1894, Nov. 23, 1895, Nov. 20, 1896, Jul. 11, 18, Nov. 3, 1897, Apr. 14, May 26, Jun. 10, Jul. 9, 18, Aug. 8, Sep. 19, Dec. 21, 1898, Apr. 16, May 15, Sep. 20, 1899; *M&SP*, May 12, Sep 8, Oct. 20, 27, Nov. 10, 1894, Jan. 1, Jul. 27, 1895, Sep. 12, 1896, Feb. 4, Mar. 4, Jun. 17, 1899; *Oregon Metal Mines Handbook*, 14-B, 133–147; Brooks and Ramp, *Gold and Silver in Oregon*, 108, 120–123; Oregon Department of

Geology and Mineral Industries, Oregon Historical Mining Information, Susanville General Report; Bureau of Land Management, General Land Office Records of Mineral Surveys and Patents, Oregon District, "Mineral Survey No. 304," patent no. 29073 (www.blm.gov/or/landrecords).

30 The discussion of the Badger Mine is based on the following: *Oregonian*, Oct. 23, 1891, Apr. 23, Jul. 9, 18, Aug. 8, Sep. 19, Dec. 21, 1898, Apr. 16, May 15, Sep. 20, 1899; *Engineering and Mining Journal*, Sep. 17, 1898; *M&SP*, Oct. 20, 1894, Feb. 26, Mar. 18, Apr. 23, Nov. 5, 26, Dec. 17, 1898, Feb. 4, Mar. 4, 25, Jun 17, 1899; Lindgren, *Gold Belt*, 706–707; *Oregon Metal Mines Handbook*, 14-B, 133–137, Brooks and Ramp, *Gold and Silver in Oregon*, 108–109, 120; Oregon Department of Geology and Mineral Industries, Oregon Historical Mining Information, Badger Group Report.

31 *Oregonian*, Sep. 19, 1898.

32 The discussion of the Granite District lode and placer mines is based on the following: Lindgren, *Gold Belt*, 677–688; Tabor, *Granite and Gold*, 17–23, 26–36; Books and Ramp, *Gold and Silver in Oregon*, 57, 62–65,77–81; *Oregon Metal Mines Handbook*, 14-B, 40–65; Brooks, *Pictorial History of Gold Mining*, 92–104; George Koch, *Lode Mines of the Central Part of the Granite Mining District, Grant County Oregon*, Oregon Department of Geology and Mineral Industries Bulletin No. 49 (Portland, OR: Department of Geology and Mineral Industries, 1959), 1–49; *Oregonian*, Nov. 20, 1896, Jan. 1, Jun. 10, Dec. 15, 1898, May 15, 1899; *M&SP*, Sep. 22, 1895, Apr. 13, Sep. 11, 1897, Jul 2, 1898, Aug. 19, Oct. 14, 1899.

33 Lindgren, *Gold Belt*, 680.

34 Lindgren, *Gold Belt*, 680–683; Tabor, *Granite and Gold*, 27–31; Brooks and Ramp, *Gold and Silver in Oregon*, 107, 117; Brooks, *Pictorial History of Gold Mining*, 82–123; *Oregon Metal Mines Handbook*, 14-B, 60–63; *Souvenir Edition*, 37; *M&SP*, Jul. 2, 1898, Aug. 19, 1899; *Engineering and Mining Journal*, Nov. 19, 1898, May 20, 1899; W. H. Tinney, *Gold Mining Machinery*, 196–197.

35 Jan. 1, 1898.

36 *The West Shore*, Feb. 1, 1885, 42–43; US Department of the Interior, Bureau of Land Management, General Land Office Records, Oregon District, Portland, Oregon, "Survey Plat Details;" Anonymous, *An Illustrated History*, 400; *Oregonian*, Sep. 19, 1898.

37 Carlos Schwantes, *The Pacific Northwest*, rev. ed. (Lincoln: University of Nebraska Press, 1996), 261–262; Johansen and Gates, *Empire of the Columbia*, 360–365; Willingham, *Starting Over*, 84–85; Peter Boag, *Pioneering Death: The Violence of Boyhood in Turn-of-the-Century Oregon* (Seattle, WA: University of Washington Press, 2022), 100–101, 105–108; US Department of Agriculture, Bureau of Statistics, *Wheat Crops of the United States, 1866-1906* (Washington DC: Government Printing Office, 1908), 35, 39; Chester Whitney Wright, "Wool-Growing and the Tariff Since 1890," *The Quarterly Journal of Economics*, Vol. 19, No. 4 (Aug. 1905), 611, 635–639; Anonymous, *An Illustrated History*, 409.

38 Quotes on 410, 175.

39 State of Oregon, Secretary of State, *Biennial Report*, 1905 (Salem, OR: State Printer, 1904), 56–57 (pullout).

40 *M&SP*, Jan. 13, 1891, Jan. 6, 1900; Brooks and Ramp, *Gold and Silver in Oregon*, 9.

CHAPTER FIVE

1 *Oregonian*. Jan. 1, 1900.

2 *Oregonian*, Jan. 1, 1900.

3 *Oregonian*, Jan. 1, 1900.

4 *Oregonian*, Jan. 1, 1900.

5 *Oregonian*, Jan. 1, 1900.

6 *Oregonian*, Jan. 1, 1900.

7 *Pacific Miner*, Sep. 1, 1902.

8 P. Donan, *Baker County Oregon, the Bonanza Land for 1900* (Portland, OR: Oregon Railroad and Navigation Company, 1900), 2–3, 15.

9 Baker City Chamber of Commerce, *The Gold Fields of Eastern Oregon* (Baker City, OR: no pub., 1900), 8.

10 *Pacific Miner*, Jul. 15, 1902.

11 *Sumpter Miner*, Dec. 24, 1902.

12 Marks, *Precious* Dust, 185; Baker City Chamber of Commerce, *The Gold Fields*, 39; US Bureau of the Census, Twelfth Census, Population Schedules (Washington, DC: microfilm publication T624), Roll 1280; *Sumpter Miner*, Jan. 3, Nov. 15, 29, 1900, Jan. 2, 1901, Jul. 30, Dec. 24, 1902, Jan. 7, 1903; *M&SP*, Dec. 1, 1900; *Pacific Miner*, Jul. 15, 1902; *Oregonian*, Mar. 11, 1900; Anonymous, "Cradle of Wealth" (Sumpter, OR: no pub., 1903).

13 *Blue Mountain Eagle*, Jun. 28, Oct. 4, 1901, Jan. 31, Jun. 20, Aug. 15, Sep. 12, Oct. 3, Nov. 7, 21, Dec. 12, 1902; *M&SP*, Mar. 8, 1902; *East Oregonian*, Aug. 22, 1902, Jul. 17, 1903, Nov. 15, 1905; the townsite of Galena was platted in 1904, *Sumpter Miner*, Aug. 24, 1904.

14 *Sumpter Miner*, Nov. 21, 1900.

15 Brooks, *Pictorial History of Mining*, 91, 121–125; Brooks and Ramp, *Gold and Silver in Oregon*, 65–67, 81–83; H. M. Parks and A. M. Swartley, *Handbook of the Mining Industry of Oregon*, Vol. 2, No. 4 (Corvallis, OR: Oregon Bureau of Mines and Geology, 1916), 20–21, 121–123, 150–151, 257; J. T. Pardee and D. F. Hewett, "Mineral Resources of the Sumpter Quadrangle," *Mineral Resources of Oregon*, Vol.1, No. 6 (Corvallis, OR: Oregon Bureau of Mines and Geology, 1914), 74–80; Arthur Swartley, "Ore Deposits of Northeastern Oregon," *The Mineral Resources of Oregon*, Vol. 1, No. 8 (Portland, OR: Oregon Bureau of Mines and Geology, 1914), 160–162; *Oregon Metal Mines Handbook*, Bulletin 14-A (Portland, OR: Oregon Department of Geology and Mineral Industries, 1939), 84–85, 87–89; Lindgren, *Gold Belt*, 645–648; Baker City Chamber of Commerce, *The Gold Fields*, 12, 15; US Department of the Interior, Bureau of Land Management, General Land Office, Historical Index, Sec. 19, 20, 21, T. 8S, R. 38E.

16 This and the next paragraph are based on Brooks, *Pictorial History of Mining*, 91, 107–109; Brooks and Ramp, *Gold and Silver in Oregon*, 59–62, 74–77; Parks and Swartley, *Handbook of Mining*, 59–65, 251–252; Pardee and Hewett, "Mineral Resources," 81–96; Swartley, "Ore Deposits," 148–160; *Oregon Metal Mines*, 14-A, 33–34, 36–41; Lindgren, *Gold Belt*, 654–655, 658-66; Baker City Chamber of Commerce, *The Gold Fields*, 15–18; *Pacific Miner*, Aug. 15, Nov.1, 1902, Jan. 1, 15, Apr.15, Jul. 15, Nov.15, 1903, Mar. 15, Apr.1, May. 1, 15, Jul. 15, Aug. 15, 1904, Mar.1, Apr.1, 15, Nov. 1, 1905; *M&SP*, Jul 6, 1901, Apr. 5, 26, May 24, Jul. 5, 12, Oct. 18, 1902, Mar 14, Aug. 8, 1903, Mar. 19, Apr. 9, Aug. 13, Dec. 3, 1904, Apr. 8, 1905, Jan. 19, 1907, Jul. 24, 1909, Apr. 23, Jul. 31, 1910; *The Engineering and Mining Journal*, Mar. 9, 1901, Jan. 5, 1905, Jul. 12, 1911; GLO, Historical Index, Sec. 28, 29, 32, T. 8S, R. 37E.

17 Quotes from Parks and Swartley, *Handbook of Mining*, 63–65.

18 Brooks, *Pictorial History of Mining*, 73, Brooks and Ramp, *Gold and Silver in Oregon*, 106–108, Parks and Swartley, *Handbook of Mining*, 29–32, 39, 187–191, 270–271; *Oregon Metal Mines*, 14-A, 52–53, 14-B, 60–63, 66–72, 88; Lindgren, *Gold Belt*, 680–685, 692–695, 700–702; Baker City Chamber of Commerce, *The Gold Fields*, 26, 28–29, 41; *Sumpter Miner*, Jul. 2, Aug. 6, 1902; *Blue Mountain Eagle*, Dec. 26, 1902; GLO, Historical Index, Sec. 35, 36, T. 9S, R. 34E, Sec. 10, T. 9S, R. 35E, Sec. 1,2, T. 10S, 35E, Sec. 10, T. 10S, 35 1/2E.

19 *Oregon Metal Mines*, 14-B, 60.

20 Quote in Baker City Chamber of Commerce, *The Gold Fields*, 26; Brooks, *Pictorial History of Mining*, 82 -84; Tabor, *Granite and Gold*, 27 -31; Pardee and Hewett, "Mineral Resources," 110–114; Parks and Swartley, *Handbook of Mining*, 188–191; *Pacific Miner*, Aug. 1, Sep. 1, Oct. 15, 1902, Jan. 15, Mar. 1, Apr. 1, Jun. 1, 1903, Jan. 15, Feb. 15, Apr.

15, May 15, Jun. 15, Aug. 1, 15, Oct. 1, 1904, Jun. 1, Aug. 15, Oct. 15, Nov. 1, 15, 1905, Jan. 1906; *The Engineering and Mining Journal*, Sep. 22, 1904, Jan. 12, 1905; *M&SP*, Jan. 20, Jun. 30, Oct. 19, Dec. 8, 1900, Jan. 19, Jul. 20, 1901, Jan. 11, Sep. 27, Oct. 18, 1902, Mar. 7, Jun. 20, 1903, May 21, 1904, Aug. 12, 1905, Apr. 11, 1908, Dec. 10, 1910.

21 Gail Throop, "Fremont Powerhouse (Granite, Oregon)," National Register of Historic Places Nomination Form (Salem, OR: State Historic Preservation Office, 1983).

22 Brooks, *Pictorial History of Mining*, 79–80; Lindgren, *Gold Belt*, 700–702; Baker City Chamber of Commerce, *The Gold Fields*, 29; Oregon Department of Geology and Mineral Industries, Historical Mining Information, Report on the Consolidated Bonanza Gold Mine; *Pacific Miner*, May 15, Aug. 15, 1904, Apr. 1, 15, May 1, 15, Jun. 1, 15, Jul. 15, Aug. 1, 15, Oct. 15, 1905, Jan. 1906; *The Engineering and Mining Journal*, May 11, 25, 1905, Feb. 9, 1907; May 3, Jul. 20, Nov. 23, 1901, Apr. 24, Oct. 18, 1902, Jun. 11, 1910.

23 *M&SP*, Jan. 25, 1902.

24 P. Donan, *Oregon, Washington, Idaho and Their Resources* (Portland, OR: The Oregon Railroad and Navigation Co., 1902), 41.

25 *Pacific Miner*, Jan. 15, 1903.

26 *Pacific Miner*, Jan. 15, 1903; for examples of eastern mining investments in eastern Oregon mines, see also *Pacific Miner*, Jul. 15, Sep. 1, Nov. 1, 1903, Jun. 1, 1904.

27 *Oregonian*, Jan. 1, 1904.

28 *Pacific Miner*, Aug. 15, 1904; *Oregonian*, Jan. 1, 1904; *M&SP*, Jan. 9, Dec. 3, 1904; *The Engineering and Mining Journal*, May 5, 1904; *Sumpter Miner*, Jan.13, 1904; O. E. Stafford, "Mineral Resources and Mining Industries of Oregon, 1903," University of Oregon *Bulletin*, Vol. 1, No. 4 (1904), 22–54.

29 *Oregon Journal*, Aug. 21, 1904.

30 *Pacific Miner*, Sep. 15, 1904.

31 *Pacific Miner*, Sep. 15, 1904.

32 *Pacific Miner*, Jun. 15, 1904.

33 *Pacific Miner*, Jul. 15, 1904.

34 *Oregonian*, Jan. 2, 1905.

35 Crane, *Gold and Silver*, 560, 562, 626.

36 Quote in *Oregonian*, Jan. 2, 1905.

37 *Pacific Miner*, Jun. 1, Jul. 15, Aug. 15, Sep. 10, 1904.

38 Tabor, *Granite and Gold*, 10–11; quote in *Sumpter Miner*, Jul. 23, 1902.

39 Hendryx File (Oregon Historical Society, microfilm), May 11, 1903.

40 *Sumpter Miner*, May 6, 1903.

41 *Pacific Miner*, Feb. 15, 1903.

42 The description and operation of the Sumpter Smelter is based on the following sources: *The Mining Investor*, May 11, 1903; *Sumpter Miner*, Apr. 9, Jul. 16, 30, Aug. 6, Dec. 10, 1902, Jan. 14, Feb. 22, Mar. 11, Apr. 1, May 6, Jul. 1, 22, Dec. 2, 1903, Jan. 6, Jun. 15, Jul. 13, Aug. 17, Sep. 7, 28, Oct. 5, 26, Nov. 16, 1904, Jan. 3, 11, 18, Feb. 1, 8, 22, 1905; *M&SP*, Jan. 28, Aug. 23, 1902, Jan. 10, Apr. 4, May 9, 1903, Sep. 24, 1904, Feb. 18, 1905, Jan. 19, Sep. 21, Nov.16, 1907, Apr. 11, 1908, Jul. 31, 1909; *Oregonian*, Jan.1, 1904, Feb. 4, 1911; *Pacific Miner*, Jul. 15, 1902, Feb. 15, Jul. 15, Aug. 15, Sep. 1, Dec. 1, 1903, Jan. 15, Apr. 1, 15, Jun 15, Sep. 15, Oct. 1, 15, Nov. 1, 15, 1904, Jan. 1, Apr. 1, May 15, Jun. 15, Jul. 1, 15, Sep. 1, Oct. 1, Nov. 15, 1905; *The Engineering and Mining Journal*, Sep. 29, 1904, Jan. 12, May 11, 1905, Mar. 10, 1906, Jan. 7, 1911; *Morning Democrat* (Baker City), Jul. 6, 1921; *Baker Herald*, Jun. 15, 1921.

43 *Pacific Miner*, Oct. 1, 1904.

44 *Oregonian*, Oct. 14, 1922, Nov. 4, 28, Dec. 31, 1923, Apr. 1, 1924, Jul. 1, Sep. 19, 1926, Jan. 10, 1927; *Blue Mountain American*, Jul. 30, 1910.

45 The discussion of Oregon's timber resources and the Sumpter Valley Railway is based on the following sources: *The Timberman*, Jan. 1904, 56–59, Jun. 1907, 44–46, Jun. 1909, 20–22; *Oregonian*, Nov. 25, 1900, Jan. 1, 1901, Jun.4, Aug. 20, Nov. 14, 1909; *Oregon Journal*, Sep. 8, 1907; Portland Chamber of Commerce, *Oregon: The Land of Opportunity* (Portland, Portland Chamber of Commerce, 1909), 35, 58, 62; William G. Robbins, *Landscapes of Promise: The Oregon Story, 1800–1940* (Seattle: University of Washington Press, 1997), 222–237; Mallory H. Ferrell, *Rails, Sagebrush, and Pine* (San Marino, CA: Golden West Books, 1967); Jim Eccles, et al., *Steaming Toward Sumpter, 1890-2002: A Brief history of the Sumpter Valley Railroad* (Sumpter Valley Railroad Restoration, 2002); Alfred Mullett and Leonard Merritt, *Sumpter Valley Railway* (Charleston, SC: Arcadia Publishing Co., 2009), 7–11; Ward Tonsfelt, "Sumpter Valley Railroad National Register District," National Register Nomination Form (Salem OR: State Historic Preservation Office, 1985), 8, 18–20; Anonymous, *Illustrated History*, 414, 451, 721; Herman Oliver, *Gold and Cattle Country* (Portland, OR: Binfords and Mort, 1961), 188–202; *Sumpter Miner*, May 30, Jun. 13, 1900, Jan. 2, Jun. 19, 1901, Apr. 1, 8, 15, Sep. 9, 1903, Jan. 6, 13, Jun. 1, Oct. 26, 1904, Feb. 1, 1905; *Pacific Miner*, Jun. 1, Nov. 15, 1903, Jun. 1, 1904, Jan. 1, Apr. 15, Jun. 15, Jul. 1, Aug. 1, Oct. 15, 1905; *The Engineering and Mining Journal*, Jan. 7, 1911; *Grant County News*, Jun. 30, 1904.

46 *Oregon Journal*, Aug. 21, 1904.

47 The material for the Badger mine story comes from the following: *Oregon Metal Mines*, 14-B, 133–137; Parks, and Swartley, *Handbook of Mining*, 19–20; Swartley, "Ore Deposits," 170–171; Oregon Department of Geology and Mineral Industries, Oregon Historical Mining Information, Report on Visit to Susanville by O. H. Hershey, Oct. 15, 1908, Badger Group Reports, and Susanville General Report; *M&SP*, Jan. 20, Sep. 8, 22, 29, 1900, Jan. 19, Jun. 22, 1901, Jan. 18, Mar. 1, 8, May 24, Jun. 21, Jul. 5, Aug. 30, Sep. 6, 1902, Jan. 10, Feb. 2, 18, May 2, 30, Jul. 4, 1903, Feb. 27, Apr. 9, Aug. 13, Sep. 3, 10, Oct. 1, 1904, Apr. 1, Jul. 15, Nov. 25, 1905; *Pacific Miner*, Jan. 1, 15, Jun. 1, 1903, Aug. 15, Nov. 1, 1904, Mar. 1, May 15, Jun. 1, Aug. 15, Sep. 15, 1905, Jan. 1906; *Oregon Journal*, Aug. 3, 1905; *The Engineering and Mining Journal*, Nov. 10, 1904, Jan. 12, Aug. 15, 1905; *Sumpter Miner*, Jan. 10, Mar. 28, Apr. 18, Jul. 4, Aug. 15, Sep. 19, 26, Oct. 3, 10, Nov. 21, 1900, Jan. 16, Feb. 6, 10, Mar. 13, 27, Apr. 21, May 22, Jun. 12, 26, Aug. 28, 1901, Jun. 25, Jul. 2, 23, Sep. 17, Oct. 29, Nov. 5, 1902, Jun. 24, 1903, Mar. 16, Jun. 1, Oct. 12, Dec. 28, 1904, Feb. 1, Mar. 15, Apr. 5, 1905; *Blue Mountain Eagle*, almost every issue between Jan. 1901 and Jan. 1903 has an article on the Badger Mine and see especially Feb. 15, Mar. 22, Jun. 7, 21, 28, Aug. 23, Oct. 4, 1901, Feb. 28, May 16, 23, Jun. 13, 20, 27, Jul. 4, 18, 25, Aug. 22, Sep. 12, Oct. 17, 31, Dec. 26, 1902, Mar. 27, Apr. 3, Jun. 26, Aug. 21, Oct. 30, 1903, May 20, 27, Aug 19, Oct. 28, Nov. 4, Dec. 23, 1904, Mar. 3, 24, May 19, Aug. 11, Nov. 24, 1905; *East Oregonian*, Dec. 12, 27, 1901, Jul. 19, 1902, May 22, 1903; *Blue Mountain American*, Oct. 5, 1901, Mar. 30, Apr. 19, 23, May 10, 31, Jun. 9, Dec. 20, 1902, Nov. 7, 1903; *Baker City Herald*, Feb. 22, 1902; *Grant County News*, Nov. 9, 1905 *Oregonian*, Jul. 17, 1927; Spence, *Mining Engineers*, 47, 276–277. In his application for a patent for the Badger Group of Claims (102 acres of mining ground), Bradley claimed that he had made improvements, which included 585 feet of shafts and 20 feet of cuts valued at $15,300; a new mill and contents, $15,000; 1,201 feet of tunnels, $12,010; shaft house and contents, $2,500; and a Blacksmith shop, $75.00; Bureau of Land Management, General Land Office Records of Mineral Surveys and Patents, Oregon District, "Mineral Survey No. 476," patent no. 39098 (www.blm.gov/or/landrecords) .

48 *Pacific Miner*, Aug. 15, 1904.

49 *East Oregonian*, May 16, 1903; each wagon was drawn by a team of four horses; see also *Sumpter Miner*, May 16, 1900.

50 *M&SP*, Oct. 18, 1902, *Pacific Miner*, Feb. 15, May 15, Aug. 15, 1904, Apr. 15, Jun. 1, 15, Jul. 1, 1905; *The Engineering and Mining Journal*, Jan. 7, 1911; *Sumpter Miner*, Jan. 3, 1905.

51 *The Engineering and Mining Journal*, Aug. 12, 1911.

52 *Pacific Miner*, Nov. 1, 1902.

53 Quoted in *Pacific Miner*, Nov. 1, 1904; on the bad effects of "wildcat" stock promoters, see also *The Engineering and Mining Journal*, Jan. 7, 1911, and *M&SP*, Feb. 28, 1903, Sep. 30, 1905, Dec. 15, 1906, Jan. 5, Feb. 9, Aug. 24, 31, Oct. 12, 26, 1907; *Blue Mountain Eagle*, Sep. 28, 1906.

54 *Oregon Journal*, Sept. 8, 1907.

55 *Oregonian*, Jan. 1, 1912.

56 The Balliet story was covered in the *Oregonian*, Oct. 28, 1900, Mar. 19, 1901, Jun. 4, Nov. 2, 1902, Dec. 10, 1903, Apr. 24, Nov. 17, 1904; *Sumpter Miner*, May 9, 1900, Jun. 12, 1901, Jun. 8, 1904; *East Oregonian*, Jan. 12, 1904; *Oregon Journal*, Feb. 25, 1905; *The Athena Press*, Aug. 25, 1905; *The Engineering and Mining Journal*, Jan. 12, 1905; *Blue Mountain Eagle*, Oct. 21, 1904, Mar. 24, 1905; O. E. Stafford, "Mineral Resources of Oregon," 54; Anonymous, *An Illustrated History*, 769–770.

57 Mar. 1, 1905.

58 Brooks, *A Pictorial History of Mining*, 116–117; for discussion of the rail link to Bourne, see *Sumpter Miner*, Jun. 19, Jul. 3, 1901, Mar. 15, 22, Apr. 5, 1905.

59 Potter, *Oregon's Golden Years*, 158–164; for an example of a *Bourne News* column, see *Sumpter Miner*, May 6, 1903; for other coverage of White, see *Sumpter Miner*, Dec. 30, 1903, Aug. 24, 1904; *Oregonian*, Aug. 23, 1904, Jul. 14, 1910, Mar. 6, 1928; *Oregon Journal*, Aug. 14, 27, 1904, Jun. 27, 1916, Jun. 1, 1941; *East Oregonian*, May 14, 1913; *Capital Journal* (Salem), Apr. 1915; *Pacific Miner*, Jun. 15, 1904.

60 Robert F. Bruner and Sean Carr, *The Panic of 1907: Lessons Learned from the Market's Perfect Storm* (New Jersey: John Wiley & Sons, 2007); www.federalreswervehistory.org/essays/panicof1907; *Oregonian*, Feb. 4, 1911; *Grant County News*, Jan. 23, 1908; *M&SP*, Jan. 5, Feb. 9, Oct. 12, 26, Nov. 9, 30, 1907, Jan. 4, 1908.

61 Quote in Richard White, *The Republic for Which It Stands: The United States During Reconstruction and the Gilded Age, 1865–1896* (New York, Oxford University Press, 2017), 609.

62 Elliot West, *Continental Reckoning: The American West in the Age of Expansion* (Lincoln, NE: University of Nebraska Press, 2023), 397.

63 Richard White, *"It's Your Misfortune and None of My Own": A History of the American West* (Norman, OK: University of Oklahoma Press, 1991), 260.

64 Benjamin Mountford and Stephen Turrnell, eds., *A Global History of Gold Rushes* (San Francisco, University of California Press, 2018), 27–29, 139–57, 166–168, 193.

65 Oregon Department of Geology and Mineral Industries, Oregon Historical Mining Information, Eastern Oregon Mining Co./ North Pole, 1895–1908; see also, *M&SP*, Dec. 3, 1904. Not all Cracker Creek "mother lode" mines were ready to give up after 1907. At the Eureka and Excelsior, Johnathan Bourne continued development work and hired San Francisco consultant F. W. Bradley (of Badger Mine fame) to solve its milling problems; *M&SP*, Jan. 19, 1907.

66 J. T. Pardee, and D. F. Hewett, "Geology and Mineral Resources of the Sumpter Quadrangle," in *Mineral Resources of Oregon*, Vol. 1, No. 6 (Portland, OR: The Oregon Bureau of Mines and Geology, 1914), 10–13, 74, 79, 82, 87, 89, 92, 110; the first experiments with dredging on the John Day River began in 1902, with the Pomeroy Dredge, which weighed 750 tons, *Pacific Miner*, Jul. 15, 1902; *Blue Mountain Eagle*, Oct. 25, 1901, Jul. 25, 1902.

67 *Oregonian*, Jan. 1, 1908.

68 *Blue Mountain Eagle*, May 21, 1909, see also, Jun. 25, 1909.

69 *M&SP*, Jan. 1, 1910.

70 *M&SP*, Feb. 5, 1910.

71 *Oregonian*, Feb. 4, 1911; for continued mining, see also *M&SP*, Dec. 29, 1906, Jan. 19, 1907, Jul. 24, 1909, Apr. 23, 30, 1910; *The Engineering and Mining Journal*, Mar. 26, Jul. 16, 1910, Jul. 1, 1911.

72 *Oregonian*, Jan. 7, 1911.

73 Frank W. Benson and Ben W. Olcott, compilers, *Oregon Blue Book* (Salem, OR: Willis S. Duniway, State Printer, 1911), 29, 41.

74 Pardee and Hewett. "Geology and Mineral Resource of the Sumpter Quadrangle, 12–13.

75 Albert Burch, "Development of Metal Mining in Oregon," *Oregon Historical Quarterly*, vol. 43 (Jun. 1942), 117.

76 US Bureau of the Census, *Mines and Quarries, 1902* (Washington, DC: Government Printing Office, 1905), 288-290; US Bureau of the Census, *Twelfth Census of the United States* (Washington, DC: Government Printing Office, 1902), *Reports*, Vol. 2 "population," pt. 2, 137; Bureau of the Census, *Thirteenth Census of the United States, Abstract of the Census, with Supplement for Oregon* (Washington, DC: Government Printing Office, 1913), 671–674.

77 The data in Table 5.5 was compiled from the *Oregonian*, January 1, 1901, 1906, and 1911 annual editions; for the assessed property values, see State of Oregon, Secretary of State, *Biennial Report for 1905*, 56 (pull-out), *1907*, 28a, *1911*, 113 (pull-out); occupations are from Schwantes, *The Pacific Northwest*, 331.

78 *Oregonian*, Sep. 19, 1898.

79 The following demographic analysis is based on the US Bureau of the Census, Twelfth Census Population Schedules (Washington, DC: Microfilm Publication No. T-623), Roll, 1347 and Thirteenth Census Population Schedules (Washington, DC: Microfilm Publication No. T-624), Roll 1280.

80 US Department of the Interior, Bureau of Land Management, GLO Records of Mineral Surveys and Patents, Oregon District (www.blm.gov/or/landrecords); Tabor, *Granite and Gold*, 10–12.

81 Wyman, *Hard Rock Epic*, 60.

82 Comparative material is from Wyman, *Hard Rock Epic*, 41–60; James, *The Roar and the Silence*, 138–139, 244–245.

CHAPTER SIX

1 Christopher Edson, "The Chinese in Eastern Oregon, 1860–1890" (master's theses, University of Oregon, 1970), 33–37; Laban Steeves, "Chinese Gold Miners of Northeastern Oregon, 1862–1900" (master's theses, University of Oregon, 1984), 40–168; Sue Fawn Chung, *In Pursuit of Gold: Chinese American Miners and Merchants in the American West* (Urbana, IL: University of Illinois Press, 2011), 54–59; Bennet Bronson and Chui-mei Ho, *Coming Home in Gold Brocade: Chinese in Early Northwest America* (Seattle, WA: Chinese in Northwest America Research Committee, 2015), 44–46, 62–63, 67; Don Hann, "Chinese Mining *Kongsi* in Eastern Oregon," *Oregon Historical Quarterly*, vol. 122 (Winter, 2021), 346; Katee Withee, "Stacked Rock Features: Archaeological Evidence of Chinese Miners on the Malheur National Forest," *Oregon Historical Quarterly*, vol. 122 (Winter, 2021), 372–373; Priscilla Wegars, The Ah Hee Diggins: Final Report of Archaeological Investigations at OR-GR-16, The Granite, Oregon "Chinese Walls" Site, 1992–1994 (Moscow, ID: University of Idaho, 1995), 10–11, 33–42; *Oregonian*, Aug. 29, 1866, Apr. 1, 1867; *Daily Mountaineer* (The Dalles), May 21, 1866.

2 Quote in Raymond, *Statistics of Mines and Mining*, House Ex. Doc. No. 10 (1871), 182; *Engineering and Mining Journal*, Feb. 28, 1871.

3 Chung, *In Pursuit of Gold*, 57; Hann, "Chinese Mining *Kongsi*," 358–359; Withee, "Stacked Rock Features," 372–373; Marks, *Precious Dust*, 299; Paul and West, *Mining Frontiers*, 144; Robert J. Swartout, Jr., "From Guangdong to Big Sky: The Chinese in Montana, 1864–1900," in *Montana: A Cultural Medley*, ed. By Robert J. Swartout, Jr. (Helena, MT: Farcountry, 2015), 98–101.

4 Mae Ngai, *The Chinese Question: The Gold Rushes and Global Politics* (New York: W. W. Norton and Company, 2021), 149.; Michael Luo, *Strangers in the Land: Exclusion,*

Belonging, and the Epic Story of the Chinese in America (New York: Doubleday, 2025), 99-114.

5 See census discussions in chapters three and five.

6 Edson, "The Chinese in Eastern Oregon," 9–10, 45. See also, Luo, *Strangers in the Land*, 26-28, 35.

7 For examples of discrimination, see Luo, *Strangers in the Land*, passim; Bronson and Ho, *Coming Home in Gold Brocade*, 43. 48–62; Chung, *In Pursuit of Gold*, 30–54; Ngai, *The Chinese Question*, passim; Beth Lew-Williams, *The Chinese Must Go* (Cambridge, MA: Harvard University, 2018), passim; *West Shore*, Mar. 1, Oct. 1, 1886; Swartout, Jr., "From Guangdong to Big Sky," 110–12; Paul and West, *Mining Frontiers*, 244–252; For a dated view of the Chinese mining experience in the West, see Marks, *Precious Dust*, 298–303.

8 *Grant County News*, Oct. 15, 1885.

9 *Grant County News*, Feb. 15, 1892.

10 *Grant County News*, Feb. 12, 1885; Anonymous, *An Illustrated History*, 436, 443; the move from Canyon City to John Day may not have completely removed the Chinese from the former, as the *Grant County News* (Nov. 12, 1898), in reporting on the losses from the devastating fire of 1898 in Canyon City wrote, "China town wiped out of existence."

11 Anonymous, *An Illustrated History*, 443.

12 Tabor, *Granite and Gold*, 45.

13 *Grant County News*, Jun. 7, 28 (quote), 1879.

14 Hann, "Chinese Mining *Kongsi*," 347; Mining Claims Conveyance Record, A-B, 1864–1881, Grant County Court House; *Grant County News*, Jun. 2, 1887, Jun. 29, 1888.

15 *Grant County News*, Oct. 9, 1880; Chung, *In Pursuit of Gold*, 69.

16 See above in chapter three and Raymond, *Statistics of Mines and Mining*, House Ex. Doc. No. 210, (1873), 212; quote in Raymond, *Statistics of Mines and Mining*, House Ex. Doc. 10 (1871), 182; see also *Oregonian*, Aug. 6, 1879; *Grant County News*, Sep. 17, Oct. 4, 1888; *Engineering and Mining Journal*, Feb. 28, 1871.

17 *Grant County News*, Oct. 4, 1888; see also *GCN*, Sep. 27, 1888.

18 Swartout, "From Guangdong to Big Sky," 112–112; Mark T. Johnson, *The Middle Kingdom under the Big Sky: A History of the Chinese Experience in Montana* (Lincoln, NE: University of Nebraska Press, 2022), 22, 192–192; Ngai, *The Chinese Question*, 45–47; Wyman, *Hard Rock Epic*, 29, 38–41, 46, 165; Paul and West, *Mining Frontiers*, 247–248.

19 *East Oregonian*, May 19, 1877.

20 *Grant County News*, Jun. 28, 1879.

21 Chung, *In Pursuit of Gold*, 72–80; Tom Adams and Christina Sweet, *On the Shelves of Kam Wah Chung* (Oregon State Heritage Site, 2013).

22 Hann, "Chinese Mining *Kongsi*," 350.

23 Anonymous, *An Illustrated History*, 442.

24 Oct. 4, 1901. In the 1900 census, he is listed as Ti Wang, aged 42, with an occupation given as laundryman. According to the census, he also owned his home.

25 Paul and West, *Mining Frontiers*, 149.

26 Quote in Hann, "Chinese Mining *Kongsi*," 350–351.

27 Quote in Trimble, *The Mining Advance Into the Inland Empire*, 231, see also 144.

28 Hann, "Chinese Mining *Kongsi*," 351.

29 Hann, "Chinese Mining *Kongsi*," 354–358; Chung, *In Pursuit of Gold*, 1–30, 59–60; Mae Ngai, *The Chinese Question*, 32–48; Johnson, *The Middle Kingdom under the Big Sky*, 4, 217n4.

30 Chung, *In Pursuit of Gold*, 57; Raymond, *Statistics of Mines and Mining*, House Ex. Doc., No. 10 (1871), 183–184; *Engineering and Mining Journal*, Feb. 28, 1871; *M&SP*, May 17, 1879; Hiatt, *Thirty-One Years in Baker County*, 84–86; Brooks, *A Pictorial History*

of Gold Mining, 45, 146; Rossiter Raymond, *Statistics of Mines and Mining in the States and Territories West of the Rocky Mountains*, 44th Cong. 1st Sess., House Ex. Doc. No. 159 (Washington, DC: Government Printing Office, 1876), 233–234. In 1908, mining expert Walter Crane, in *Gold and Silver* (p. 358), accurately described Chinese mining techniques but failed to explain how they acquired them; see also, Mike Higgins and Les Tipton, *Ditch Walkers and Water Wars: The Life and Times of the Eldorado Ditch in the Gold Fields of Eastern Oregon* (Caldwell, ID: Blue Pine Publishing Co., 2013), 12–35, 57–108 100–101, 129–134, 153–171, 196–198, 239.

31 Hann, "Chinese Mining *Kongsi*," 358.

32 Hann, "Chinese Mining *Kongsi*," 359; Chung, *In Pursuit of Gold*, 59, 69; US Bureau of the Census, Products of Industry Schedule of the Nineth Census, Grant County, Oregon, manuscript at Oregon State Library; *Sumpter Miner*, Aug. 15, 1900, and May 22, 1901.

33 State Board of Agriculture, *The Resources of the State of Oregon* (Salem OR, 1899), 28.

34 The following discussion of eastern Oregon mining archaeology is based on Withee, "Stacked Rock Features," 368–388; Wegars, *The Ah Hee Diggins* (1995); Hann, "Chinese Mining *Kongsi*, 360–363; Chelsea Rose and Katie Johnson, "Chinese Mining in the Blue Mountains: Life and Work Along the Middle Fork of the John Day River, Malheur National Forest," SOULA Report No. 2021.11 (2022), 12–17, 27–32, 154–160; see also, https://oregon-chinese-diaspora-project-sou.hub.argis.com/.

35 Hann, "Chinese Mining *Kongsi*," 361.

36 Withee, "Stacked Rock Features," 371.

37 Hann, "Chinese Mining *Kongsi*," 359; Chung, *In Pursuit of Gold*, 72–81; Jeffery Barlow and Christine Richardson, *China Doctor of John Day* (Portland, OR: Binford and Mort, 1979), 4–97; Sally Donovan and Sarah Griffith, "Kam Wah Chung Company Building," National Historic Landmark Nomination, John Day, Oregon, 2005; for an example of Kam Wah Chung Company mining investment, see *Grant County News*, Jul. 5, 1887.

38 Horatio C. Burchard, *Report of the Director of the Mint upon the Statistics of the Production of the Precious Metals in the United States for 1882* (Washington, DC: Government Printing Office, 1883), 180; see chapter three above for other reports of the amount of gold recovered by the Chinese during the 1880s in Elk Creek District.

39 *Grant County News*, Sep. 9, 1886.

40 Raymond, *Statistics of Mines and Mining*, House Ex. Doc., No. 10 (1871), 177.

41 Edward G. Jones, *The Oregonian's Handbook of the Pacific Northwest* (Portland, OR: The Oregonian Publishing Co., 1894), 70.

42 *Souvenir Edition*, 46.

43 P. Donan, *Baker County Oregon, The Bonanza Land for 1900*, 15.

44 *Sumpter Miner*, Mar. 30, 1904.

45 Anonymous, *An Illustrated History*, 757.

46 Anonymous, *An Illustrated History*, 219.

47 Chung, *In Pursuit of Gold*, 72; see also, H. Gregory Nokes, *Massacred for Gold: The Chinese in Hells Canyon* (Corvallis, OR: Oregon State University Press, 2009); Luo, *Strangers in the Land*, 319-321.

48 According to the federal census records, John Day went from five Chinese merchants in 1900 to just one in 1910. In Susanville, the number of Chinese merchants in 1910 increased to three from just two in 1900. The 1905 state census of Oregon recorded 105 Chinese in Baker County and twenty-five in Grant County; State of Oregon, *Biennial Report of the Secretary of State*, Enumeration of Inhabitants (Salem, OR: State Printer, 1907), 106. D. H. Leon, for example, was listed in the Susanville federal census of 1900 as a miner, in the 1910 census as a merchant, and in 1920 census as a farm laborer. He was still at Susanville in 1927, where he apparently operated a gas station, *Oregonian*, Jul. 17, 1927; US Bureau of the Census, Twelfth Census, Population Schedules (Microfilm

Publication Number T623), roll 1347, Thirteenth Census, Population Schedules (Microfilm Publication Number T624), roll 1280.

49 Oliver, *Gold and Cattle Country*, 158–160.

CHAPTER SEVEN

1 Mountford and Tuffnell, eds. *A Global History of Gold Rushes*, 4.

2 William L. Silber, *The Story of Silver: How the White Metal Shaped America and the Modern World* (Princeton, NJ: Princeton University Press, 2019), 12–23; Matthew Hart, *Gold: The Race for the World's Most Seductive Metal* (New York: Simon & Schuster, 2013), 49–50.

3 Silber, *The Story of Silver*, 24–25; White, *The Republic for Which It Stands*, 370–371.

4 White, *The Republic for Which It Stands*, 371.

5 Hart, *Gold*, 50–56; Richard Bensel, *The Political Economy of American Industrialization, 1877–1900* (New York: Cambridge University Press, 2000), 349–456.

6 The state-wide discussion of the Populist Party in Oregon in the following paragraphs draws on the following books and article: Schwantes, *The Pacific Northwest*, 261–270; Gordon B. Dodds, *The American Northwest: A History of Oregon and Washington* (Arlington Heights, IL: The Forum Press, 1986), 144–152, 178–184; Robbins, *Oregon, This Storied Land*, 78–82, 91; Robert D. Johnston, *The Radical Middle Class: Populist Democracy and the Question of Capitalism in Progressive Era Portland, Oregon* (Princeton, NJ: Princeton University Press, 2003), 121–124, 127–131; MacColl, *Merchants, Money and Power*, 319–329; MacColl, *The Shaping of a City*, 200–212; Walter M. Pierce, *Oregon Cattleman, Governor, Congressman*, ed., and expanded by Arthur Bone (Portland, OR: Oregon Historical Society Press, 1981), 37, 39–41, 418; Thomas C. McClintock, "Seth Lewelling, William S. U'Ren and the Birth of the Oregon Progressive Movement," in *Oregon Historical Quarterly*, vol. 68 (Sep. 1967), 197–220. For a perceptive analysis of the political economy of Populism and its consequences, see Bensel, *The Political Economy of American Industrialization*, 101–288.

7 Anonymous, *An Illustrated History*, 488–489, 769, 773; George Turnbull, *The History of Oregon Newspapers* (Portland, OR: Binfords and Mort, 1939), 354–355. Unfortunately, few, if any, copies of these papers have survived.

8 Anonymous, *An Illustrated History*, 190.

9 Anonymous, *An Illustrated History*, 191.

10 Anonymous, *An Illustrated History*, 421–425.

11 See sources in note 6 above for the discussion of fraud and bribery.

12 The discussion in this and the following two paragraphs is based on Brooks and Ramp, *Gold and Silver*, 4–8, 10, 18–21, 27–28, 44, 47, 49, 89–90, 129; S. H. Lorain, *Gold Mining and Milling in Northeastern Oregon*, US Bureau of Mines Information Circular, No. 7015 (Washington, DC, 1938); George S. Koch, Jr., *Lode Mines of the Central Part of the Granite Mining District, Grant County, Oregon*, Oregon Department of Mineral Industries, Bulletin No. 49 (Portland, OR, 1959); Hawley, *Gold Dredging*, 1–34; Anonymous, *An Illustrated History*, 413, 761; Brooks, *A Pictorial History of Gold Mining*, 42–44, 56, 59.

13 Brooks and Ramp, *Gold and Silver*, 5.

14 Brooks and Ramp, *Gold and* Silver, 28.

15 This and the next paragraph are based on Baken, *The Mining Law of 1872*, 116–187; Coming, *Mining Explained*, 65–72; Oregon Department of Geology and Mineral Industries, Mineral Land Regulations and Reclamation—Regulations and Statutes, accessed at www.oregon.gov/dogami/mlrr/Pages/regulations.

16 Quote in Baken, *The Mining Law of 1872*, 127.

17 Curtis, *Gambling on Ore*, 18.

Index